Classical Mechanics and Geometry

Classical Mechanics and Geometry

Si Li
Department of Mathematical Sciences
Tsinghua University

International Press of Boston, Inc. · Somerville, Massachusetts, U.S.A.

Classical Mechanics and Geometry

by Si Li
Department of Mathematical Sciences, Tsinghua University

Somerville, Massachusetts, U.S.A.
www.intlpress.com

ISBN (print): 978-1-57146-454-5
ISBN (online): 978-1-57146-455-2

Printed in China, September 2023

Contents

Preface

In April 2021, Qiuzhen College (求真书院) was newly established at Tsinghua University under the leadership of Professor Shing-Tung Yau. It homes the distinguished elite mathematics program in China starting in 2021: the "Yau Mathematical Sciences Leaders Program" (丘成桐数学科学领军人才培养计划). This program puts strong emphasis on basic sciences related to mathematics in a broad sense. Though majored in mathematics, students in this program are required to study fundamental theoretical physics such as classical mechanics, electromagnetism, quantum mechanics, and statistical mechanics, in order to understand global perspectives of theoretical sciences. It is an exciting challenge both for students and for instructors.

This preliminary note is written for the course "Classical Mechanics" that I lectured at Qiuzhen College in the fall semester of 2022. It is to explain key physics ingredients of Lagrangian and Hamiltonian mechanics, as well as their connections with modern geometric development. We put heavy emphasis on different faces of concrete examples in order to understand the bridge between mathematics and physics. Examples such at Toda lattice and Calogero-Moser System are still active research topics nowadays in areas of integrable system, representation theory and mathematical physics. A large part of this note relies on the beautiful books of "Mechanics" by Landau-Lifshitz, and "Mathematical Methods of Classical Mechanics" by Arnol'd, which themselves show different faces of this classical subject. Other useful resources that we consulted are listed at the end of this note.

I would like to thank Yang Peng (杨鹏) and Wang Jinyi (王进一), who have done amazing jobs of teaching assistant for this course. An early version of this note was typed by Yang Peng, including all those beautiful figures that are better arts than my blackboard drawings. I want to thank Ding Xu Zhihan (丁徐祉晗) and Liu Jiuhe (刘九和) for their help on careful proofreading of this note, as well as their important roles of being excellent students for the whole semester. I want to thank my colleague Zhou Jie (周杰), the collaboration and discussion with whom in this year have kept my brain fresh during the preparation of this note. Special thank goes to Cheng Ziyu (程子钰) from office of Teaching Affairs at Qiuzhen College, whose tremendous help has saved me alive from heavy administrative service to finish this note.

Jan 1, 2023

Chapter 1

Lagrangian Mechanics

1.1 Principle of Least Action

1.1.1 Newtonian Mechanics

Recall **Newton's Second Law**

$$\vec{F} = m\vec{a}.$$

Consider a particle moving in $\mathbb{R}^n$ at time $t \in \mathbb{R}$ with position $\vec{q}(t) \in \mathbb{R}^n$. Let us describe this motion as a map

$$\begin{aligned} \vec{q} : \mathbb{R} &\longrightarrow \mathbb{R}^n \\ t &\longmapsto \vec{q}(t) = (q_1(t), \cdots, q_n(t)) \end{aligned}$$

We have

- **velocity**: $\vec{v} = \dot{\vec{q}} = \frac{d\vec{q}}{dt}$
- **acceleration**: $\vec{a} = \dot{\vec{v}} = \frac{d^2\vec{q}}{dt^2}$

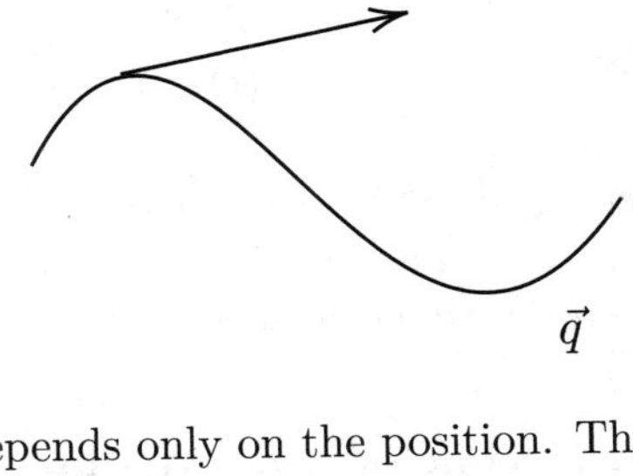

Assume the force $\vec{F}$ depends only on the position. Then

$$m\ddot{\vec{q}} = \vec{F}(q(t)),$$

which has a unique solution if we fix initial value $(\vec{q}(t_0), \dot{\vec{q}}(t_0))$ at some time t_0.

Definition 1.1.1. Define the **kinetic energy** of the motion

$$K = \frac{1}{2}m\vec{v}^2 = \frac{1}{2}m\left(\frac{d\vec{q}}{dt}\right)^2 = \frac{1}{2}\sum_i m\left(\frac{dq_i}{dt}\right)^2.$$

We will mainly consider conservative forces, in which case we can write[1] $\vec{F} = -\nabla V$ for some function $V : \mathbb{R}^n \to \mathbb{R}$ called the **potential**[2]. The **potential energy** of a particle is, by definition, the potential at the location of the particle. The total energy is defined by

$$E = K + V.$$

Proposition 1.1.2. *The total energy is conserved, i.e., independent of time along the motion.*

Proof.

$$\begin{aligned}\frac{dE}{dt} &= \frac{d}{dt}K + \frac{d}{dt}V = \frac{d}{dt}\left(\frac{1}{2}m\dot{\vec{q}}\cdot\dot{\vec{q}}\right) + \sum_i \frac{dq_i}{dt}\frac{\partial}{\partial q_i}V \\ &= m\ddot{\vec{q}}\cdot\dot{\vec{q}} + \sum_i \nabla V\cdot\dot{\vec{q}} = (\vec{F} + \nabla V)\cdot\dot{\vec{q}} = 0.\end{aligned}$$ □

We will study systematically conservation laws in Section 1.3 via Noether's Theorem.

1.1.2 Action Functional

Definition 1.1.3. Define the **Lagrangian** of the system of motion by

$$\mathscr{L} = K - V.$$

We assume $\mathscr{L} = \mathscr{L}(q, \dot{q}, t)$ depends on $\vec{q}$, $\dot{\vec{q}}$ and t[3]. For any path

$$\vec{q}(t) : [t_0, t_1] \longrightarrow \mathbb{R}^n,$$

we define its **action functional** by

$$S[\vec{q}(t)] = \int_{t_0}^{t_1} \mathscr{L}(q(t), \dot{q}(t), t)dt.$$

As we will see, the trajectories of classical particles are stationary points of the system's action functional on the path space. Often though not always, the action is minimized for classical trajectories, then this is the least action. Such law is generally called the

Principle of Least Action.

[1] In components, $\vec{F} = (F_1, \cdots, F_n) = -(\partial_1 V, \cdots, \partial_n V)$.

[2] The potential in $\mathbb{R}^n$ is determined up to a constant.

[3] Theory also applies to cases when $\mathscr{L}$ contains higher time derivatives of q such as $\ddot{q}$, etc.

Intuitively, since $K+V=E$ is conserved, minimizing $\mathscr{L}=K-V$ prefers "smaller K" and "bigger V". For example, consider a ball rolling down a hill. It starts with bigger V, and somehow prefers staying there for longer time. You can use similar idea to think about a projectile motion. The particle becomes more lazy at the top to form a parabola trajectory.

1.1.3 Principle of Least Time

Before we develop the principle of least action for mechanics, we discuss some examples of such principle in physics. We first consider Fermat's Principle of least time: when a ray of light goes from one point to another, it chooses its path that takes the least time. The connection between mechanics and optics will be illustrated in Section 2.6.

Example 1.1.4 (Reflection). *A ray of light approaches a polished surface and reflects back.*

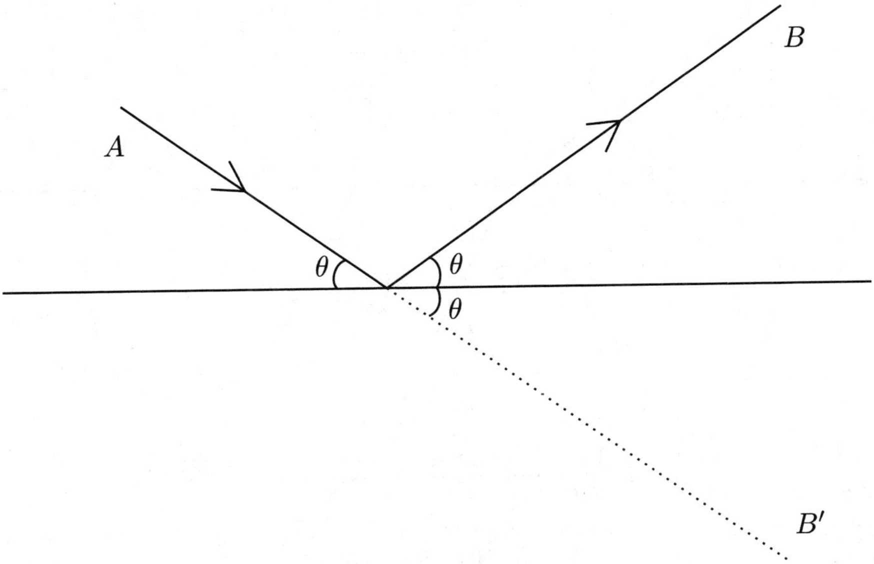

Example 1.1.5 (Snell's law). *Let n_i be the index of refraction of medium i. In the medium, the light travels with velocity $v_i = c/n_i$. Snell's law says*

$$n_1 \sin\theta_1 = n_2 \sin\theta_2.$$

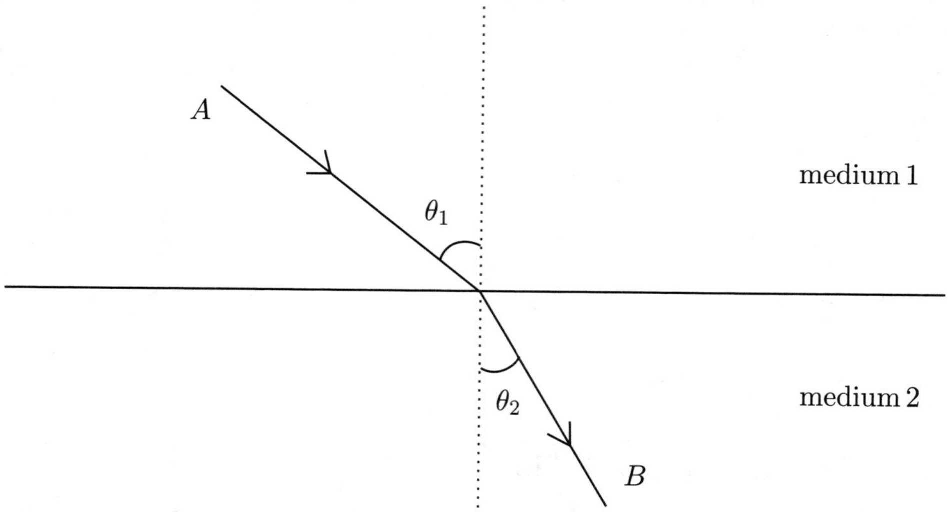

Let us see how this follows from the principle of least time. The travel time is

$$T = \frac{n_1 l_1}{c} + \frac{n_2 l_2}{c}$$

where l_i is the travel distance in medium i.

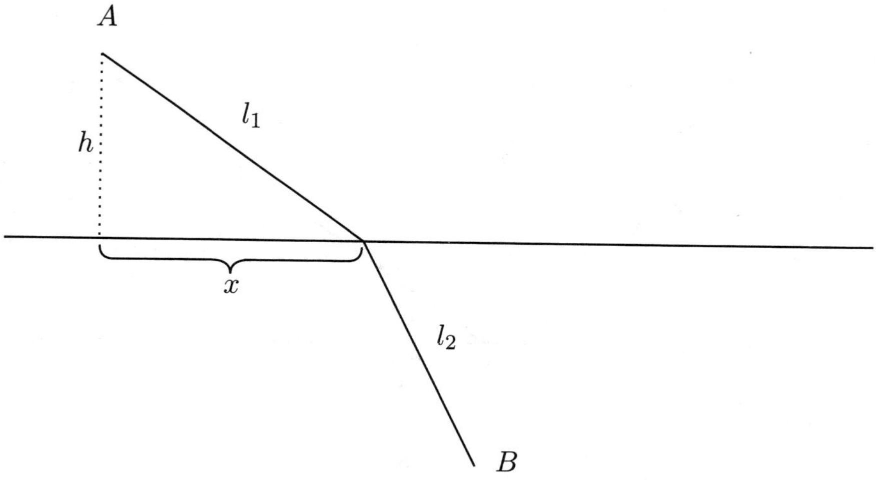

Let h be the distance of A to the interface, and x be the coordinate on the interface as in the figure. Then $l_1 = \sqrt{h^2 + x^2}$. We have

$$\frac{dl_1}{dx} = \frac{x}{\sqrt{h^2 + x^2}} = \sin\theta_1.$$

Similarly consider l_2 as a function of x and find $\frac{dl_2}{dx} = -\sin\theta_2$. Minimizing the time T asks for

$$\frac{d}{dx}T = n_1\frac{dl_1}{dx} + n_2\frac{dl_2}{dx} = 0,$$

from which we conclude

$$n_1 \sin\theta_1 = n_2 \sin\theta_2.$$

1.1.4 Principle of Minimum Energy

As another example, we consider static equilibrium of mechanical object which is achieved by minimizing its energy.

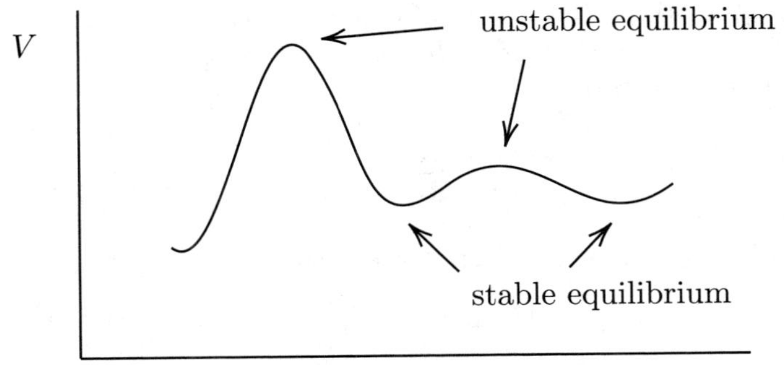

equilibrium: $dV = 0$

Example 1.1.6 (Archimedes law of the lever). *The potential of the lever in the figure is*

$$V = m_1gh_1 + m_2gh_2.$$

Here $h_1 = l_1 \sin\theta$, $h_2 = -l_2 \sin\theta$. In equilibrium,

$$K = 0, \qquad \left.\frac{dV}{d\theta}\right|_{\theta=0} = m_1gl_1 - m_2gl_2 = 0.$$

We get Archimedes law

$$m_1l_1 = m_2l_2.$$

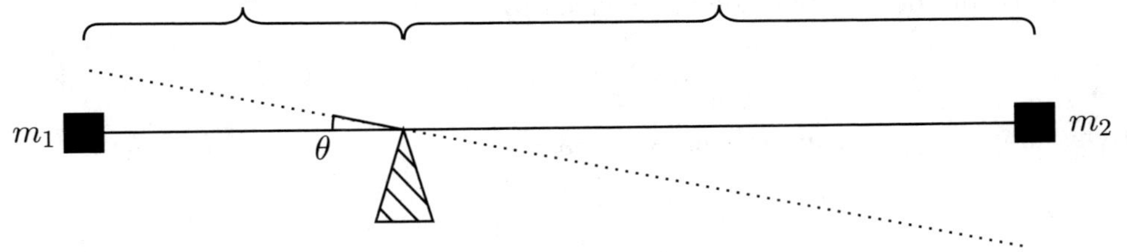

1.2 Euler-Lagrange Equation

1.2.1 Calculus of Variations

For simplicity, let us first assume the 1-dim case with position

$$\begin{aligned} q : [t_0, t_1] &\longrightarrow \mathbb{R} \\ t &\longmapsto q(t) \end{aligned}$$

Assume $\mathscr{L} = \mathscr{L}(q, \dot{q}, t)$ only depends on q, $\dot{q}$, t but not higher $\ddot{q}$, etc.

Theorem 1.2.1. *Assume $x : [t_0, t_1] \to \mathbb{R}$ is a smooth path that extremizes the action functional $S = \int_{t_0}^{t_1} \mathscr{L}\, dt$ for all possible smooth paths $q : [t_0, t_1] \to \mathbb{R}$ with the same end points: $q(t_0) = x(t_0)$, $q(t_1) = x(t_1)$. Assume the Lagrangian $\mathscr{L} = \mathscr{L}(q(t), \dot{q}(t), t)$ only depends on $q(t)$, its time derivative $\dot{q}(t)$ and time t. Then $x(t)$ satisfies the* ***Euler-Lagrange Equation***

$$\frac{d}{dt}\left(\frac{\partial \mathscr{L}}{\partial \dot{x}}\right) = \frac{\partial \mathscr{L}}{\partial x}.$$

Proof. Let $x(t)$ be such an extremizer. For any smooth map

$$\gamma : [t_0, t_1] \longrightarrow \mathbb{R}, \qquad \gamma(t_0) = \gamma(t_1) = 0$$

and any small number s, the path

$$\widetilde{x}(t) = x(t) + s\gamma(t)$$

has the same end points with x:

$$\widetilde{x}(t_i) = x(t_i) + s\gamma(t_i) = x(t_i), \qquad i = 0, 1.$$

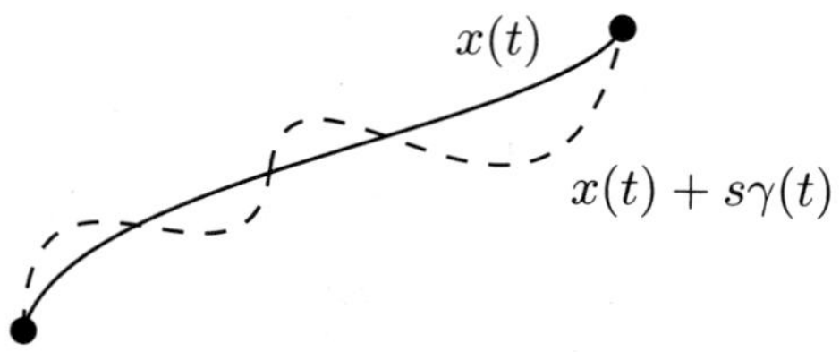

Let us consider the function of s defined by

$$f(s) := S[x + s\gamma].$$

By assumption, f has an extremal value at $s = 0$, so

$$\left.\frac{df}{ds}\right|_{s=0} = 0.$$

On the other hand,

$$f(s) = \int_{t_0}^{t_1} \mathscr{L}(x + s\gamma, \dot{x} + s\dot{\gamma}, t)\, dt.$$

By chain rule,

$$\begin{aligned}
\left.\frac{df}{ds}\right|_{s=0} &= \int_{t_0}^{t_1} \left(\frac{\partial \mathscr{L}}{\partial x}\gamma + \frac{\partial \mathscr{L}}{\partial \dot{x}}\dot{\gamma} \right) dt \\
&= \int_{t_0}^{t_1} \left(\frac{\partial \mathscr{L}}{\partial x}\gamma + \frac{d}{dt}\left(\frac{\partial \mathscr{L}}{\partial \dot{x}}\gamma \right) - \left(\frac{d}{dt}\frac{\partial \mathscr{L}}{\partial \dot{x}} \right)\gamma \right) dt \\
&= \int_{t_0}^{t_1} \left(\frac{\partial \mathscr{L}}{\partial x} - \frac{d}{dt}\frac{\partial \mathscr{L}}{\partial \dot{x}} \right)\gamma\, dt + \left.\frac{\partial \mathscr{L}}{\partial \dot{x}}\gamma\right|_{t_0}^{t_1} \\
&= \int_{t_0}^{t_1} \left(\frac{\partial \mathscr{L}}{\partial x} - \frac{d}{dt}\frac{\partial \mathscr{L}}{\partial \dot{x}} \right)\gamma\, dt
\end{aligned}$$

which is zero for arbitrary choice of γ. It follows that

$$\frac{\partial \mathscr{L}}{\partial x} - \frac{d}{dt}\left(\frac{\partial \mathscr{L}}{\partial \dot{x}} \right) = 0. \qquad \square$$

Remark. $\frac{d}{dt}\left(\frac{\partial \mathscr{L}}{\partial \dot{x}}\right)$ is the total dependence on t involving x, $\dot{x}$, and t. Explicitly, for

$$\frac{\partial \mathscr{L}}{\partial \dot{x}} = \frac{\partial \mathscr{L}}{\partial \dot{x}}(x, \dot{x}, t),$$

we have

$$\frac{d}{dt}\left(\frac{\partial \mathscr{L}}{\partial \dot{x}} \right) = \left(\dot{x}\frac{\partial}{\partial x} + \ddot{x}\frac{\partial}{\partial \dot{x}} + \frac{\partial}{\partial t} \right)\left(\frac{\partial \mathscr{L}}{\partial \dot{x}} \right).$$

The above calculation generalizes to the n-dim case

$$q : [t_0, t_1] \longrightarrow \mathbb{R}^n.$$

The corresponding extremal path satisfies the **Euler-Lagrange Equation**

$$\frac{\partial \mathscr{L}}{\partial q_i} - \frac{d}{dt}\left(\frac{\partial \mathscr{L}}{\partial \dot{q}_i} \right) = 0 \qquad \text{for } i = 1, \cdots, n.$$

1.2.2 Comparison with Calculus

In calculus, given a function $f(x)$, we have a meaning of total differential

$$df = \frac{\partial f}{\partial x} dx,$$

where dx means the infinitesimal variation of x.

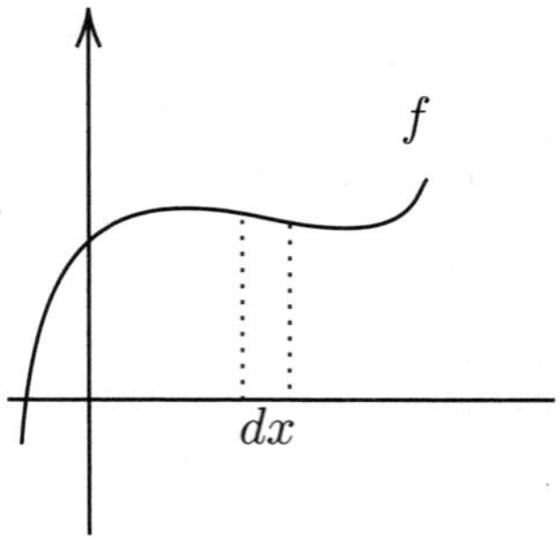

We can do the similar thing in the path space

$$\mathfrak{X} = \{\text{smooth paths } q : [t_0, t_1] \to \mathbb{R}^n\}.$$

The action functional can be viewed as a function on the infinite dimensional space $\mathfrak{X}$

$$\begin{aligned} S : \mathfrak{X} &\longrightarrow \mathbb{R} \\ q(t) &\longmapsto S[q] \end{aligned}$$

We can write δq for an infinitesimal variation. Then the total differential δS computes

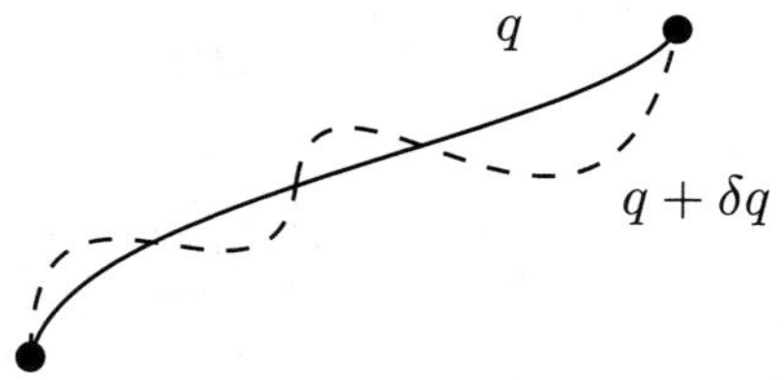

$$\begin{aligned} \delta S = \delta \int_{t_0}^{t_1} \mathscr{L}(q, \dot{q}, t)\, dt &= \int_{t_0}^{t_1} \left(\frac{\partial \mathscr{L}}{\partial q} \delta q + \frac{\partial \mathscr{L}}{\partial \dot{q}} \delta \dot{q} \right) dt \\ &= \int_{t_0}^{t_1} \left(\frac{\partial \mathscr{L}}{\partial q} - \frac{d}{dt} \frac{\partial \mathscr{L}}{\partial \dot{q}} \right) \delta q\, dt + \frac{\partial \mathscr{L}}{\partial \dot{q}} \delta q \Big|_{t_0}^{t_1}. \end{aligned}$$

For an extremal path q, $\delta S = 0$ for any δq that vanishes at fixed endpoints. Therefore

$$\frac{\partial \mathscr{L}}{\partial q} - \frac{d}{dt}\left(\frac{\partial \mathscr{L}}{\partial \dot{q}} \right) = 0,$$

which gives a simple derivation of the Euler-Lagrange equation. The rigorous mathematical aspect of this derivation is precisely the content of Theorem 1.2.1, under the price of wording. We will try to strike a balance between rigor and art mostly in this book, and leave the readers to figure out the relevant content of their preference.

1.2.3 Examples

Example 1.2.2. *Let us consider a conservative force $\vec{F}$ with potential V. Consider a particle moving under such a force. The Lagrangian is*

$$\mathscr{L} = \sum_i \frac{1}{2} m \dot{q}_i^2 - V(q).$$

We calculate

$$\frac{\partial \mathscr{L}}{\partial q_i} = -\frac{\partial V}{\partial q_i}, \qquad \frac{\partial \mathscr{L}}{\partial \dot{q}_i} = m\dot{q}_i, \qquad \frac{d}{dt}\left(\frac{\partial \mathscr{L}}{\partial \dot{q}_i}\right) = m\ddot{q}_i.$$

The Euler-Lagrange Equation reads

$$m\ddot{q}_i = -\frac{\partial V}{\partial q_i} = F_i.$$

This is Newton's 2nd law. If $V = 0$, the particle is free. The Euler-Lagrange equation becomes

$$m\ddot{q}_i = 0.$$

Hence the particle moves in a straight line with constant velocity.

Example 1.2.3 (Lorentz force). *We consider a particle of charge Q moving with velocity $\vec{v}$ in an electric field $\vec{E}$ and a magnetic field $\vec{B}$. The particle experiences a force*

$$\vec{F} = Q(\vec{E} + \vec{v} \times \vec{B}),$$

which is called Lorentz force. Both $\vec{E}$ and $\vec{B}$ can be expressed in terms of

- *a scalar potential $\phi(t, x_1, x_2, x_3)$*
- *a vector potential $\vec{A}(t, x_1, x_2, x_3) = (A_1, A_2, A_3)$*

by

$$\begin{cases} \vec{E} = -\nabla\phi - \dfrac{\partial \vec{A}}{\partial t} \\ \vec{B} = \nabla \times \vec{A} \end{cases}$$

Let us denote the position vector by $\vec{x} = (x_1, x_2, x_3)$. The Lagrangian is

$$\mathscr{L} = \frac{1}{2} m \dot{\vec{x}}^2 - Q\phi + Q\vec{A} \cdot \dot{\vec{x}} = \frac{1}{2} m(\dot{x}_1^2 + \dot{x}_2^2 + \dot{x}_3^2) - Q\phi + Q(A_1\dot{x}_1 + A_2\dot{x}_2 + A_3\dot{x}_3).$$

Here the potential

$$V = Q(\phi - \vec{A} \cdot \dot{\vec{x}})$$

depends on $\dot{\vec{x}}$. Let us derive the Euler-Lagrange Equation:

$$\frac{\partial \mathscr{L}}{\partial x_i} = -Q\frac{\partial \phi}{\partial x_i} + Q\frac{\partial \vec{A}}{\partial x_i} \cdot \dot{\vec{x}},$$

$$\frac{d}{dt}\left(\frac{\partial \mathscr{L}}{\partial \dot{x}_i}\right) = \frac{d}{dt}(m\dot{x}_i + QA_i) = m\ddot{x}_i + Q\sum_j \frac{dx_j}{dt}\frac{\partial A_i}{\partial x_j} + Q\frac{\partial A_i}{\partial t}$$
$$= m\ddot{x}_i + Q\sum_j \frac{\partial A_i}{\partial x_j}\dot{x}_j + Q\frac{\partial A_i}{\partial t}.$$

The Euler-Lagrange Equation

$$\frac{\partial \mathscr{L}}{\partial x_i} = \frac{d}{dt}\left(\frac{\partial \mathscr{L}}{\partial \dot{x}_i}\right)$$

reads

$$m\ddot{x}_i = -Q\frac{\partial \phi}{\partial x_i} + Q\frac{\partial \vec{A}}{\partial x_i}\cdot \dot{\vec{x}} - Q\sum_j \frac{\partial A_i}{\partial x_j}\dot{x}_j - Q\frac{\partial A_i}{\partial t}.$$

The electro-magnetic force is

$$F_i = -Q\left(\frac{\partial \phi}{\partial x_i} + \frac{\partial A_i}{\partial t}\right) + Q\sum_j\left(\frac{\partial A_j}{\partial x_i} - \frac{\partial A_i}{\partial x_j}\right)\dot{x}_j.$$

The first term is the component of $Q\vec{E}$ with

$$\vec{E} = -\nabla\phi - \frac{\partial \vec{A}}{\partial t}.$$

The second term is the component of $Q\dot{\vec{x}} \times \vec{B}$ with

$$\vec{B} = \nabla \times \vec{A}.$$

So we find the Lorentz force

$$\vec{F} = Q\vec{E} + Q\vec{v}\times\vec{B}.$$

Remark. The potential can be encoded into a 1-form

$$-\phi dt + A_1 dx_1 + A_2 dx_2 + A_3 dx_3.$$

Computing components of its curvature 2-form, we get the fields $\vec{E}$ and $\vec{B}$. The potential part of the action functional can be viewed as the integration of this 1-form over a trajectory of a particle viewed as a path $(t, \vec{x}(t))$ in the space-time $\mathbb{R}^{3,1}$.

Example 1.2.4 (Spring with gravity). *Suppose the spring is massless with elastic coefficient k. We know*

$$K = \frac{1}{2}m\dot{x}^2, \qquad V = \frac{1}{2}kx^2 + mgh = \frac{1}{2}kx^2 - mgx.$$

The Lagrangian is

$$\mathscr{L} = \frac{1}{2}m\dot{x}^2 - \frac{1}{2}kx^2 + mgx.$$

The Euler-Lagrange Equation

$$\frac{\partial \mathscr{L}}{\partial x} - \frac{d}{dt}\frac{\partial \mathscr{L}}{\partial \dot{x}} = 0$$

reads

$$-kx + mg - m\ddot{x} = 0,$$

or

$$m\ddot{x} = mg - kx.$$

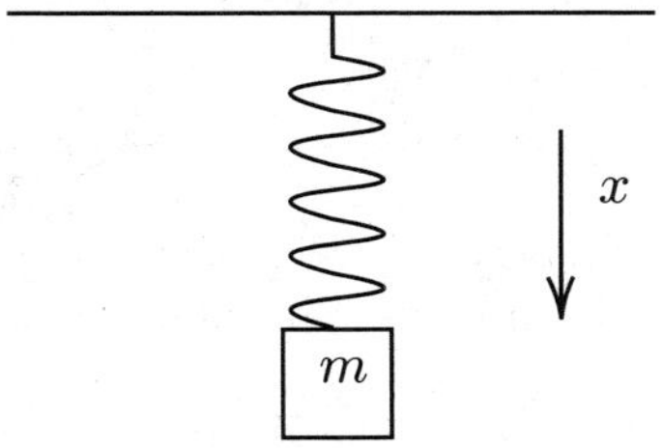

Example 1.2.5 (The Atwood Machine). *Consider the ideal Atwood machine consisting of two objects of mass $m_1 > m_2$ connected by an inextensible massless string of length l over a pulley. Assume there is no friction.*

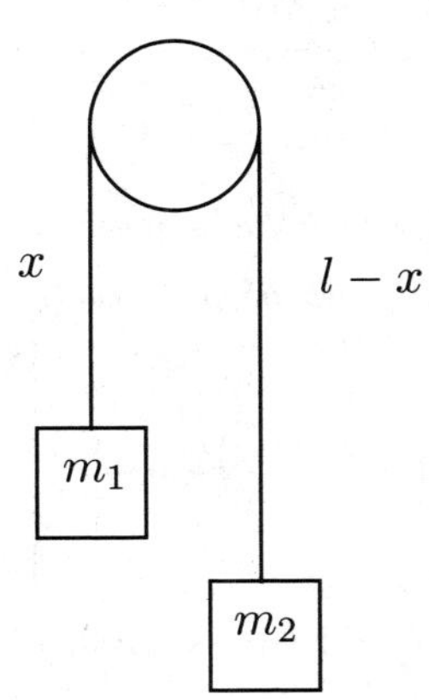

Then

$$\begin{aligned}
K &= \frac{1}{2}m_1\dot{x}^2 + \frac{1}{2}m_2\dot{(l-x)}^2 = \frac{1}{2}(m_1 + m_2)\dot{x}^2, \\
V &= m_1 g(-x) + m_2 g(-(l-x)) = -m_1 gx - m_2 g(l-x), \\
\mathscr{L} &= K - V = \frac{1}{2}(m_1 + m_2)\dot{x}^2 + m_1 gx + m_2 g(l-x).
\end{aligned}$$

The Euler-Lagrange equation

$$\frac{d}{dt}\left(\frac{\partial \mathscr{L}}{\partial \dot{x}}\right) = \frac{\partial \mathscr{L}}{\partial x}$$

reads

$$(m_1 + m_2)\ddot{x} = (m_1 - m_2)g,$$

or

$$\ddot{x} = \frac{m_1 - m_2}{m_1 + m_2}g.$$

This is equivalent to a falling object with gravitational acceleration $\frac{m_1 - m_2}{m_1 + m_2}g$.

Example 1.2.6 (Bead on a rotating rod). *Consider a bead of mass* m *sliding on a rod rotating in a horizontal plane with constant frequency* ω. *Assume there is no friction.*

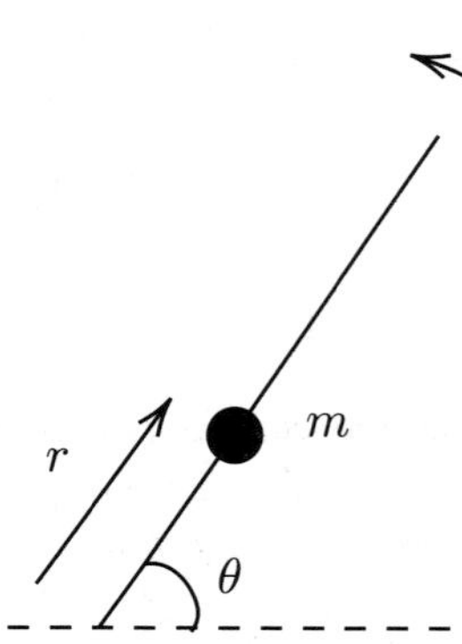

Then

$$\theta(t) = \omega t.$$

At time t, *we parametrize the position of bead on the plane by*

$$\begin{cases} x(t) = r(t)\cos\theta(t) \\ y(t) = r(t)\sin\theta(t) \end{cases}$$

with velocity

$$\begin{cases} \dot{x}(t) = \dot{r}\cos\theta - (r\sin\theta)\dot{\theta} \\ \dot{y}(t) = \dot{r}\sin\theta + (r\cos\theta)\dot{\theta} \end{cases}$$

The Kinetic energy is

$$\begin{aligned} K &= \frac{1}{2}m\left(\dot{x}^2 + \dot{y}^2\right) = \frac{1}{2}m\left[(\dot{r}\cos\theta - (r\sin\theta)\dot{\theta})^2 + (\dot{r}\sin\theta + (r\cos\theta)\dot{\theta})^2\right] \\ &= \frac{1}{2}m(\dot{r}^2\cos^2\theta - 2r\dot{r}\cos\theta\sin\theta\dot{\theta} + r^2\sin^2\theta\dot{\theta}^2 + \dot{r}^2\sin^2\theta + 2r\dot{r}\cos\theta\sin\theta\dot{\theta} + r^2\cos^2\theta\dot{\theta}^2) \\ &= \frac{1}{2}m(\dot{r}^2 + r^2\dot{\theta}^2) = \frac{1}{2}m(\dot{r}^2 + r^2\omega^2). \end{aligned}$$

The potential of the system is

$$V = 0.$$

So the Lagrangian is simply

$$\mathscr{L} = \frac{1}{2}m(\dot{r}^2 + r^2\omega^2).$$

Viewed as an 1*-dim mechanical problem along the radius* r*, this is equivalent to having a potential* $-\frac{1}{2}mr^2\omega^2$ *and force* $F = mr\omega^2$*. The Euler-Lagrange equation reads*

$$m\ddot{r} = mr\omega^2, \qquad \text{or equivalently,} \qquad \ddot{r} = \omega^2 r.$$

This is solved by

$$r(t) = ae^{\omega t} + be^{-\omega t},$$

where a, b *are constants. Assume the bead sits at* $r = r_0 > 0$ *at time* $t = 0$*. The initial condition is* $r(0) = r_0$, $\dot{r}(0) = 0$*. So we can solve* a, b *to get*

$$r(t) = \frac{1}{2}r_0(e^{\omega t} + e^{-\omega t}).$$

We find $r(t) \to +\infty$ *when* $t \to +\infty$*, as expected.*

Example 1.2.7 (Disk pulled by falling mass)**.** *Consider a disk of mass* m_1 *pulled across a table by a falling object of mass* m_2*. Assume there is no friction.*

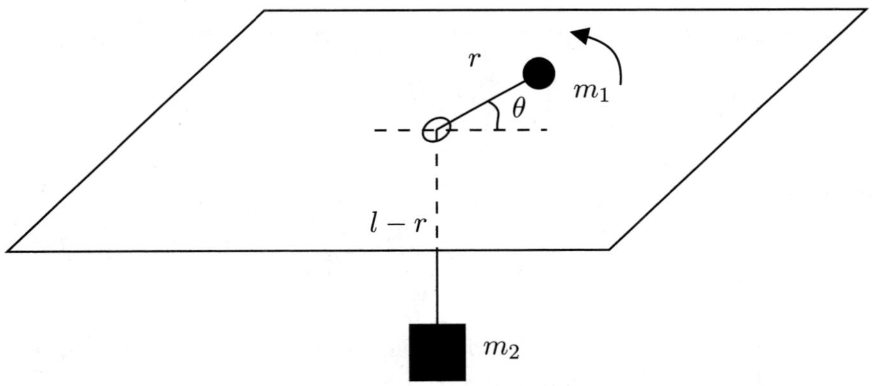

The kinetic energy of m_1 *is computed as in the previous example*

$$\frac{1}{2}m_1(\dot{r}^2 + r^2\dot{\theta}^2).$$

The kinetic energy of m_2 *is*

$$\frac{1}{2}m_2\overset{\cdot}{(l - r)}^2 = \frac{1}{2}m_2\dot{r}^2.$$

The gravitational potential is

$$V = -m_2g(l - r) = m_2g(r - l).$$

Therefore the Lagrangian is

$$\mathscr{L} = \frac{1}{2}m_1(\dot{r}^2 + r^2\dot{\theta}^2) + \frac{1}{2}m_2\dot{r}^2 - m_2 g(r - l).$$

The Euler-Lagrange equation

$$\begin{cases} \frac{d}{dt}\left(\frac{\partial \mathscr{L}}{\partial \dot{r}}\right) = \frac{\partial \mathscr{L}}{\partial r} \\ \frac{d}{dt}\left(\frac{\partial \mathscr{L}}{\partial \dot{\theta}}\right) = \frac{\partial \mathscr{L}}{\partial \theta} \end{cases}$$

reads

$$\begin{cases} (m_1 + m_2)\ddot{r} = m_1 r\dot{\theta}^2 - m_2 g \\ \frac{d}{dt}(m_1 r^2 \dot{\theta}) = 0 \end{cases}$$

From the second equation we get

$$J := m_1 r^2 \dot{\theta} = constant.$$

As we will see, the constant J is precisely the angular momentum. Plugging

$$\dot{\theta} = \frac{J}{m_1 r^2}$$

into the first equation, we find

$$(m_1 + m_2)\ddot{r} = \frac{J^2}{m_1 r^3} - m_2 g.$$

Effectively, this can be viewed as a 1-dim problem along r, where the disk feels a force $\frac{J^2}{m_1 r^3} - m_2 g$ with potential (by integrating)

$$\frac{J^2}{2m_1 r^2} + m_2 g r.$$

The stable equilibrium is at

$$\frac{J^2}{m_1 r_*^3} - m_2 g = 0.$$

We find

$$r_* = \left(\frac{J^2}{m_1 m_2 g}\right)^{1/3}.$$

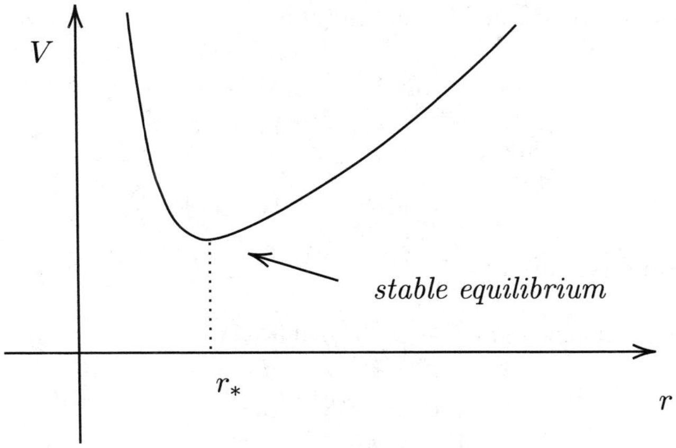

At this point, the disk rotates with frequency

$$\dot{\theta} = \frac{J}{m_1 r_*^2} = \left(\frac{m_2^2 g^2}{m_1 J}\right)^{1/3} .$$

Otherwise we would find orbits like

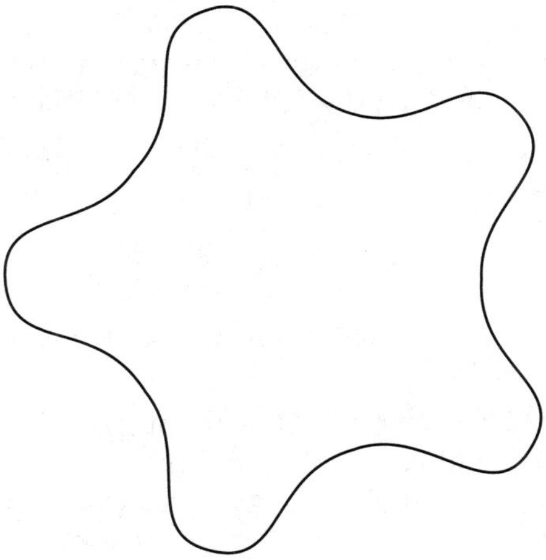

1.2.4 Plane Motion with Central Force

Now we generalize a bit about the above two examples, and consider some general properties of the motion on the plane with central force. The position is

$$\vec{r} = (x, y) = (r\cos\theta, r\sin\theta)$$

and the potential is

$$U = U(r),$$

which only depends on r. The kinetic energy is

$$K = \frac{1}{2}m\dot{\vec{r}}^2 = \frac{1}{2}m(\dot{r}^2 + r^2\dot{\theta}^2).$$

The Lagrangian is

$$\mathscr{L} = \frac{1}{2}m(\dot{r}^2 + r^2\dot{\theta}^2) - U(r).$$

The Euler-Lagrange equation

$$\begin{cases} \dfrac{d}{dt}\left(\dfrac{\partial \mathscr{L}}{\partial \dot{r}}\right) = \dfrac{\partial \mathscr{L}}{\partial r} \\ \dfrac{d}{dt}\left(\dfrac{\partial \mathscr{L}}{\partial \dot{\theta}}\right) = \dfrac{\partial \mathscr{L}}{\partial \theta} \end{cases}$$

reads

$$\begin{cases} m\ddot{r} = mr\dot{\theta}^2 - \dfrac{\partial U}{\partial r} \\ \dfrac{d}{dt}(mr^2\dot{\theta}) = 0 \end{cases}$$

Let $J = mr^2\dot{\theta}$ which is a constant. We find

$$\dot{\theta} = \frac{J}{mr^2}, \qquad m\ddot{r} = \frac{J^2}{mr^3} - \frac{\partial U}{\partial r} = -\frac{\partial}{\partial r}\left(\frac{J^2}{2mr^2} + U\right).$$

Therefore for a motion of a particle on the plane with central force with potential $U(r)$, it is effectively equivalent to a 1-dim problem along the radius with potential

$$V(r) = \frac{J^2}{2mr^2} + U(r).$$

The angular moves with $\dot{\theta} = \frac{J}{mr^2}$.

Let us explore the meaning of J. The angular momentum is defined to be $m\vec{r} \times \dot{\vec{r}}$. It points toward the normal direction, with length

$$m(x\dot{y} - y\dot{x}) = m(r\cos\theta(\dot{r}\sin\theta + r\cos\theta\dot{\theta}) - r\sin\theta(\dot{r}\cos\theta - r\sin\theta\dot{\theta})) = mr^2\dot{\theta} = J.$$

We see that J is precisely the angular momentum.

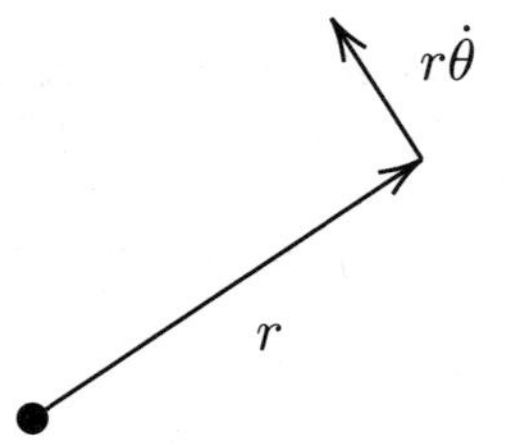

Now we consider the 1-dim problem with

$$\mathscr{L} = \frac{1}{2}m\dot{r}^2 - V(r).$$

The total energy $E = \frac{1}{2}m\dot{r}^2 + V(r)$ is a constant. We find

$$\frac{dr}{dt} = \sqrt{\frac{2(E - V(r))}{m}}, \qquad dt = \sqrt{\frac{m}{2(E - V(r))}}dr$$

$$\Longrightarrow \qquad t = \sqrt{\frac{m}{2}} \int \frac{dr}{\sqrt{E - V(r)}}.$$

On the other hand, we have

$$\frac{d\theta}{dt} = \frac{J}{mr^2}, \qquad d\theta = \frac{J}{mr^2}dt = \sqrt{\frac{1}{2m}} \frac{Jdr}{r^2\sqrt{E - V(r)}}$$

$$\Longrightarrow \qquad \theta = \sqrt{\frac{1}{2m}} \int \frac{Jdr}{r^2\sqrt{E - V(r)}}.$$

Remark. The two conserved quantities E, J allows us to solve the equation of motion completely. This is an example of a general phenomenon of integrability as we will discuss in Chapter 4. It also helps to understand the motion quantitatively. For example, consider the case

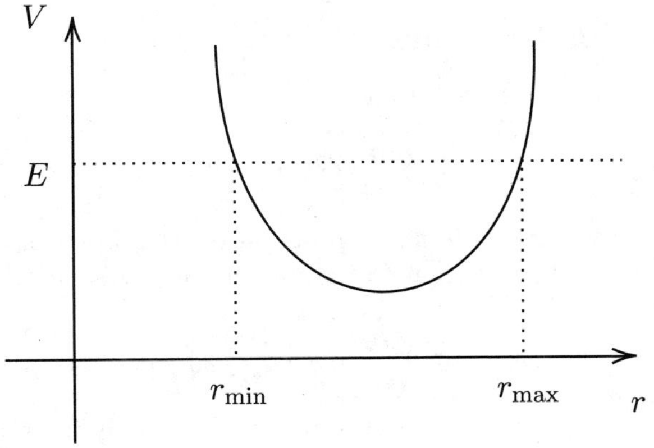

The inequality

$$V(r) \leq E$$

implies

$$r_{\text{min}} \leq r \leq r_{\text{max}}.$$

So the motion will be bounded for such potential.

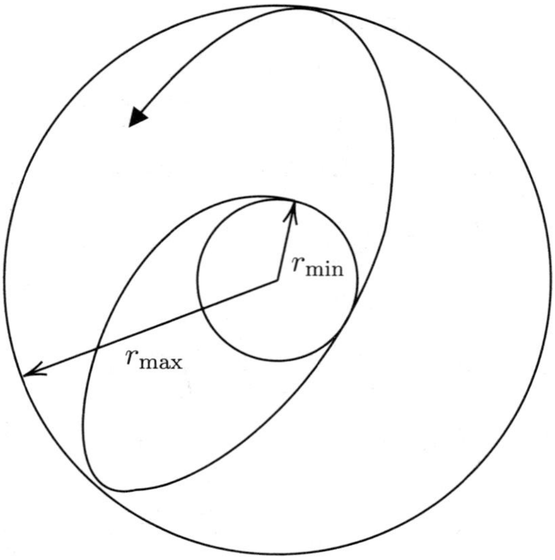

1.3 Noether's Theorem

Noether's Theorem characterizes the principle

Continuous symmetry	$\Longrightarrow$	Conservation law

We explore this principle in this section.

1.3.1 Infinitesimal Symmetry

Let $q : \mathbb{R} \to \mathbb{R}^n$ and consider an action

$$S = \int_{\mathbb{R}} \mathscr{L}(q, \dot{q}, t)\, dt.$$

Here we work with $t \in \mathbb{R}$. S is not a function on the space of all paths, since it may diverge. However, the variation δS makes sense for those δq which have compact support:

$$\delta S = \int_{\mathbb{R}} \left(\frac{\partial \mathscr{L}}{\partial q} \delta q + \frac{\partial \mathscr{L}}{\partial \dot{q}} \delta \dot{q} \right) dt.$$

Require $\delta S = 0$ for all δq with compact support, then apply integration by part and we get

$$\delta S = \int_{\mathbb{R}} \left(\frac{\partial \mathscr{L}}{\partial q} - \frac{d}{dt} \frac{\partial \mathscr{L}}{\partial \dot{q}} \right) \delta q \, dt = 0, \qquad \forall \delta q \text{ compactly supported.}$$

This still leads to the Euler-Lagrange equation

$$\frac{\partial \mathscr{L}}{\partial q} - \frac{d}{dt} \frac{\partial \mathscr{L}}{\partial \dot{q}} = 0.$$

This can be viewed as an alternative way to describe variation with fixed endpoints.

Definition 1.3.1. Let ϵ denote an infinitesimal number[4]. An infinitesimal variation of the form

$$\delta_\epsilon q = \epsilon X(q, \dot{q}, t)$$

is called a **infinitesimal symmetry** of S if the variation of the Lagrangian is a total time derivative

$$\begin{aligned}\delta_\epsilon \mathscr{L} &= \delta_\epsilon q \frac{\partial \mathscr{L}}{\partial q} + (\dot{\delta_\epsilon q}) \frac{\partial \mathscr{L}}{\partial \dot{q}} = \epsilon \left(X \frac{\partial \mathscr{L}}{\partial q} + \dot{X} \frac{\partial \mathscr{L}}{\partial \dot{q}} \right) \\ &= \epsilon \frac{d}{dt} l \qquad \text{for some } l = l(q, \dot{q}, t).\end{aligned}$$

Here the total time derivative is $\frac{dl}{dt} = \dot{q}\frac{\partial l}{\partial q} + \ddot{q}\frac{\partial l}{\partial \dot{q}} + \frac{\partial l}{\partial t}$. In this case, $\delta_\epsilon S = \int \epsilon \frac{d}{dt} l = 0$.

Example 1.3.2. *Let $\mathscr{L} = \frac{1}{2}m\dot{q}^2 - V(q)$. Consider the infinitesimal time translation $t \to t + \epsilon$ which induces a variation (using $\epsilon^2 = 0$)*

$$\delta_\epsilon q = q(t + \epsilon) - q(t) = \epsilon \dot{q}.$$

The corresponding variation for $\dot{q} = \frac{d}{dt}(q)$ when q varies is

$$\delta_\epsilon \dot{q} = \frac{d}{dt}(\delta_\epsilon q) = \epsilon \ddot{q}.$$

Then

$$\delta_\epsilon \mathscr{L} = \epsilon m \dot{q}\ddot{q} - \epsilon \dot{q} \partial_q V = \epsilon \frac{d}{dt} \underbrace{\left(\frac{1}{2} m \dot{q}^2 - V \right)}_{=l}.$$

On the other hand,

$$S = \int \left(\frac{1}{2} m \dot{q}^2 - V(q) \right) dt$$

is obviously invariant under the transformation

$$S[q(t)] = S[q(t + a)]$$

for any $a \in \mathbb{R}$. Going to the first order, we have $\delta_\epsilon S = 0$ as expected.

1.3.2 Noether's Theorem

Assume the infinitesimal transformation

$$\delta_\epsilon q = \epsilon X$$

[4]If you are not confident about infinitesimal number, you can equivalently treat ϵ as an ordinary variable and keep only the first order in expressions of ϵ in all calculations.

is an infinitesimal symmetry of S. Let us consider a more general variation

$$\delta_{\epsilon(t)} q = \epsilon(t) X$$

by promoting $\epsilon = \epsilon(t)$ to be time dependent. This can be viewed as a family of infinitesimal numbers parametrized by t. Equivalently, you can treat this as a function $\epsilon(t)$ on t and keep only the first order in expressions of $\epsilon(t)$ in all calculations. The rigorous formulation can be done as in the proof of Theorem 1.2.1, under the price of wording. We will use this concise formulation as that in Section 1.2.2, and leave it to readers to think more about it.

Then in general we will find (keeping only the first order in expressions of $\epsilon(t)$)

$$\delta_{\epsilon(t)} \mathscr{L} = \epsilon(t) \frac{d}{dt} l + \dot{\epsilon}(t)(\cdots) + \ddot{\epsilon}(t)(\cdots) + \cdots$$

which will reduce to $\delta_\epsilon \mathscr{L} = \epsilon \frac{d}{dt} l$ for some $l = l(q, \dot{q}, t)$ when $\epsilon(t) = \epsilon$ is a constant (so $\dot{\epsilon}(t) = 0$), hence an infinitesimal symmetry. Using integration by part, we can write it uniquely as

$$\delta_{\epsilon(t)} S = \int \epsilon(t) \left(\frac{d}{dt} \mathcal{J} \right) dt$$

for some expression $\mathcal{J}$. In fact, if

$$\delta_{\epsilon(t)} \mathscr{L} = \epsilon(t) \frac{d}{dt} l + \sum_{k>0} \frac{d^k \epsilon}{dt^k} \phi_k,$$

then

$$\mathcal{J} = l + \sum_{k>0} (-1)^k \left(\frac{d}{dt} \right)^{k-1} \phi_k.$$

Now we consider a path $q(t)$ that satisfies the Euler-Lagrange equation. Then

$$\delta S|_{q(t)} = 0$$

for any compactly supported variation δq. This implies

$$\int \epsilon(t) \frac{d\mathcal{J}}{dt}\bigg|_{q(t)} dt = 0$$

for any infinitesimal variation $\epsilon(t)$ with compact support. It follows that we must have

$$\frac{d\mathcal{J}}{dt} = 0$$

if $q(t)$ satisfies the Euler-Lagrange equation, i.e., $\mathcal{J}$ is a conserved quantity! This proves the following Noether's theorem.

Theorem 1.3.3. *Suppose that $\delta_\epsilon q = \epsilon X$ is an infinitesimal symmetry of the action S. Let $\delta_{\epsilon(t)} q = \epsilon(t)X$. Assume*

$$\delta_{\epsilon(t)} S = \int \epsilon(t) \frac{d\mathcal{J}}{dt}\, dt.$$

Then

$$\left.\frac{d\mathcal{J}}{dt}\right|_q = 0$$

if q satisfies the Euler-Lagrange equation.

Now let us derive an explicit expression for $\mathcal{J}$[5]. Assume

$$\delta_\epsilon \mathscr{L} = \epsilon \frac{d}{dt} l.$$

Then

$$\begin{aligned}\delta_{\epsilon(t)}\mathscr{L} = \delta_{\epsilon(t)}\mathscr{L}(q,\dot{q},t) &= \frac{\partial \mathscr{L}}{\partial q}\cdot \epsilon(t) X + \frac{\partial \mathscr{L}}{\partial \dot{q}} \cdot \frac{d}{dt}(\epsilon(t)X) \\ &= \epsilon(t)\left(\frac{\partial \mathscr{L}}{\partial q} X + \frac{\partial \mathscr{L}}{\partial \dot{q}} \dot{X}\right) + \dot{\epsilon}(t)\frac{\partial \mathscr{L}}{\partial \dot{q}} X = \epsilon(t)\frac{d}{dt} l + \dot{\epsilon}(t) \frac{\partial \mathscr{L}}{\partial \dot{q}} X,\end{aligned}$$

$$\delta_{\epsilon(t)} S = \delta_{\epsilon(t)} \int \mathscr{L}\, dt = \int \epsilon(t)\left(\frac{d}{dt} l\right) dt + \dot{\epsilon}(t)\frac{\partial \mathscr{L}}{\partial \dot{q}} X\, dt = \int \epsilon(t) \frac{d}{dt}\left(l - \frac{\partial \mathscr{L}}{\partial \dot{q}} X\right) dt.$$

So

$$\mathcal{J} = l - \frac{\partial \mathscr{L}}{\partial \dot{q}} X$$

is conserved.

Theorem 1.3.4. *Assume $\delta_\epsilon q = \epsilon X$ is an infinitesimal symmetry of $S = \int \mathscr{L}\, dt$[6]. Assume $\delta_\epsilon \mathscr{L} = \epsilon \frac{d}{dt} l$. Then*

$$\left.\frac{d}{dt}\left(\frac{\partial \mathscr{L}}{\partial \dot{q}} X - l\right)\right|_q = 0$$

if $q(t)$ satisfies the Euler-Lagrange equation. In other words, $\frac{\partial \mathscr{L}}{\partial \dot{q}} X - l$ is a conserved quantity.

[5]Here we assume $\mathscr{L} = \mathscr{L}(q, \dot{q}, t)$ does not rely on higher derivatives of q. In general, for a Lagrangian $\mathscr{L} = \mathscr{L}(t, q, \dot{q}, \ddot{q}, \cdots)$, the reader may be willing to generalize the calculation by himself.

[6]As before, we assume $\mathscr{L} = \mathscr{L}(t, q, \dot{q})$ does not rely on higher derivatives of q.

1.3.3 Energy, Momentum, and Angular Momentum

Example 1.3.5 (Time translation symmetry)**.** *The action is*

$$S = \int \left(\frac{1}{2} m\dot{q}^2 - V(q)\right) dt.$$

As in Example 1.3.2, the infinitesimal time translation symmetry is

$$\delta_\epsilon(q) = \epsilon \dot{q}.$$

So $X = \dot{q}$, $l = \mathscr{L}$ as shown above. Then

$$\frac{\partial \mathscr{L}}{\partial \dot{q}} X - l = m\dot{q}^2 - \left(\frac{1}{2} m\dot{q}^2 - V(q)\right) = K + V$$

is conserved if q satisfies the equation $m\ddot{q} = -\nabla V$. $E = K + V$ is precisely the energy!

Time translation symmetry	$\Longrightarrow$	Energy conservation

Example 1.3.6 (Space translation symmetry)**.** *Assume the Lagrangian has the property that*

$$\frac{\partial \mathscr{L}}{\partial q_i} = 0 \qquad \text{for some } q_i.$$

Such a q_i is called a cyclic coordinate. Then the infinitesimal transformation

$$\delta_\epsilon q_i = \epsilon, \qquad \text{and} \qquad \delta_\epsilon q_j = 0 \qquad \text{for } j \neq i$$

is a symmetry of S, where

$$\delta_\epsilon \mathscr{L} = \delta_\epsilon \mathscr{L}(q, \dot{q}, t) = 0.$$

Then $p_i = \frac{\partial \mathscr{L}}{\partial \dot{q}_i}$ is conserved. This is the "momentum conservation".

One such example is the free particle:

$$\mathscr{L} = \frac{1}{2} m\dot{q}^2, \qquad p = \frac{\partial \mathscr{L}}{\partial \dot{q}} = m\dot{q}.$$

Another example is a charged particle in electric-magnetic field with

$$\frac{\partial \phi}{\partial x_i} = \frac{\partial \vec{A}}{\partial x_i} = 0.$$

The Lagrangian is

$$\mathscr{L} = \frac{1}{2} m\dot{\vec{x}}^2 - Q\phi + Q\vec{A} \cdot \dot{\vec{x}}$$

from which we can derive the conserved momentum

$$p_i = \frac{\partial \mathscr{L}}{\partial \dot{x}_i} = m\dot{x}_i + QA_i.$$

We have observed an interesting shifting by QA_i for conserved momentum. In summary

Space translation symmetry	$\Longrightarrow$	Momentum conservation

Example 1.3.7 (Space rotation symmetry). *Let $M(n)$ denote all $n \times n$ real matrices. Let*

$$SO(n) = \{A \in M(n) \mid A^t A = I, \det A = 1\},$$
$$so(n) = \{X \in M(n) \mid X^t = -X, \operatorname{Tr} X = 0\}.$$

$SO(n)$ is the group of orientation preserving orthogonal transformations on $\mathbb{R}^n$. $so(n)$ is the Lie algebra of $SO(n)$, representing first order transformations. In fact, for $X \in so(n)$, its exponential e^{tX} gives a one-parameter family of elements in $SO(n)$.

The infinitesimal rotation transformation along $X \in so(n)$ is

$$\delta_\epsilon \vec{q} = \epsilon X \cdot \vec{q},$$

or in components,

$$\delta_\epsilon q_i = \epsilon \sum_j X_{ij} q_j.$$

Assume the Lagrangian is rotational invariant. For example,

$$\mathscr{L} = \frac{1}{2} m \dot{\vec{q}}^2 - V(q^2),$$

where the potential V only depends on the length of $\vec{q}$. Then the conserved quantity is

$$\sum_i \sum_j \frac{\partial \mathscr{L}}{\partial \dot{q}_i} X_{ij} q_j = m \sum_{i,j} X_{ij} \dot{q}_i q_j.$$

For each $i < j$, consider the special matrix X with $X_{ij} = -X_{ji} = -1$ and other components zero. The corresponding conserved quantity is

$$m(q_i \dot{q}_j - q_j \dot{q}_i).$$

In $n = 3$ so the space is $\mathbb{R}^3$, they are components of the vector

$$m\vec{q} \times \dot{\vec{q}}$$

which is precisely the angular momentum. We find

Space rotation symmetry	$\Longrightarrow$	Angular momentum conservation

1.4 Kepler Problem

We study motions in $\mathbb{R}^3$ with central conservative forces. Important examples are gravitational and electrostatic forces, which obey the "inverse square law".

Let $\vec{r} = (x_1, x_2, x_3)$ denote the position vector in $\mathbb{R}^3$. $r = \sqrt{x_1^2 + x_2^2 + x_3^2}$ is the length. For central conservative forces, the potential $U = U(r)$ only depends on r. The force

$$F = -\nabla U = -\partial_r U \nabla r = -\partial_r U \frac{\vec{r}}{r}$$

points toward or away from the origin.

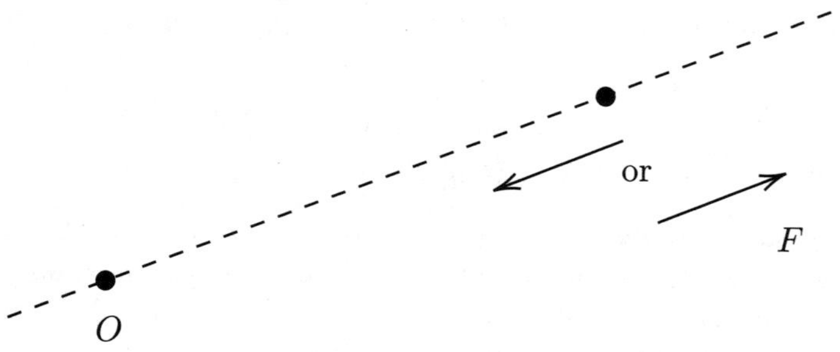

The Lagrangian is

$$\mathscr{L} = \frac{1}{2} m \dot{\vec{r}}^2 - U(r).$$

The angular momentum

$$\vec{J} = m\vec{r} \times \dot{\vec{r}}$$

is conserved. Observe that

$$\vec{J} \cdot \vec{r} = \vec{J} \cdot \dot{\vec{r}} = 0.$$

Since the direction of $\vec{J}$ is fixed, the motion of the particle is confined to the plane containing the initial position and velocity[7].

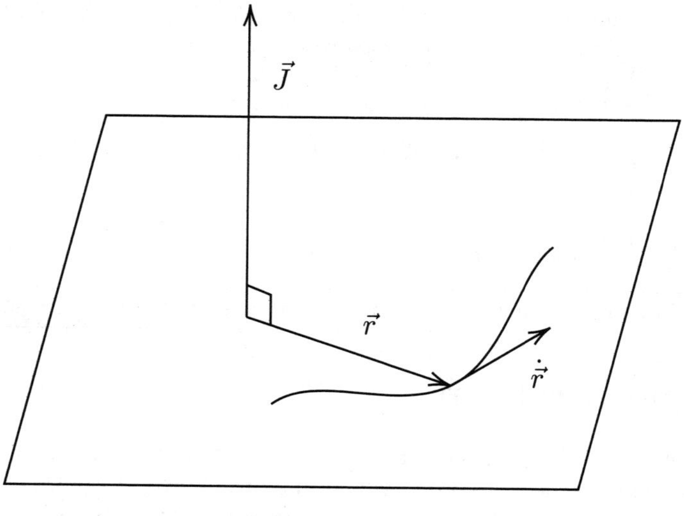

This will reduce the problem to a 2-dim motion with central conservative forces, and further reduces to a 1-dim problem with an effective potential as we have discussed in Section 1.2.4.

[7] If $\vec{J} = 0$, the motion will be confined to a line.

1.4.1 Harmonic Oscillator

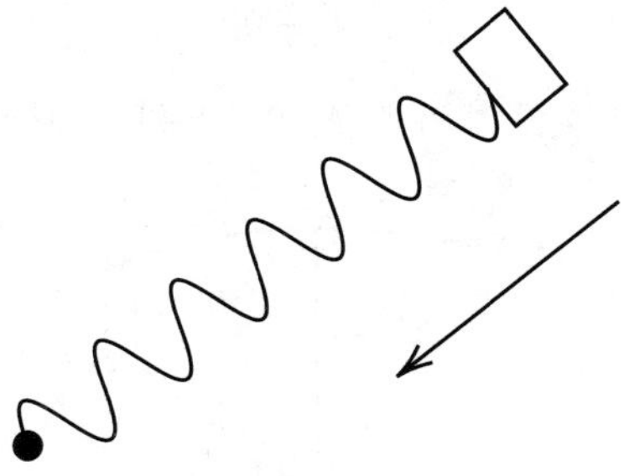

Let us first consider the classic example of harmonic oscillator. The central force is

$$\vec{F} = -k\vec{r}, \qquad k > 0.$$

The potential is

$$U(r) = \frac{1}{2}kr^2.$$

The Lagrangian is

$$\mathscr{L} = \frac{1}{2}m\dot{\vec{r}}^2 - \frac{1}{2}kr^2.$$

The equation of motion is

$$m\ddot{\vec{r}} = -k\vec{r}.$$

In term of components, this is

$$\begin{cases} m\ddot{x}_1 = -kx_1 \\ m\ddot{x}_2 = -kx_2 \\ m\ddot{x}_3 = -kx_3 \end{cases}$$

The general solution is

$$\begin{cases} x_1(t) = \widetilde{a}_1 \cos \omega t + \widetilde{b}_1 \sin \omega t \\ x_2(t) = \widetilde{a}_2 \cos \omega t + \widetilde{b}_2 \sin \omega t \\ x_3(t) = \widetilde{a}_3 \cos \omega t + \widetilde{b}_3 \sin \omega t \end{cases}$$

where $\widetilde{a}_i$, $\widetilde{b}_i$ are constants, $\omega = \sqrt{k/m}$. We can write in vectors

$$\vec{r}(t) = \vec{\widetilde{a}} \cos \omega t + \vec{\widetilde{b}} \sin \omega t.$$

By a suitable choice of a shift θ_0, we can write

$$\vec{r}(t) = \vec{a} \cos(\omega t - \theta_0) + \vec{b} \sin(\omega t - \theta_0)$$

for another two vectors $\vec{a}$, $\vec{b}$ such that they are perpendicular:

$$\vec{a} \cdot \vec{b} = 0.$$

We see the orbit of motion is an ellipse with center at the origin and semi-axes $||a||$ and $||b||$.

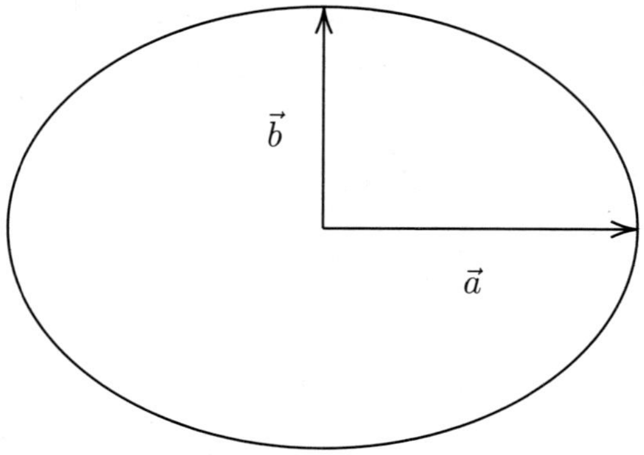

Another way to find the semi-axes is to use the effective potential in one dimension:

$$V(r) = \frac{J^2}{2mr^2} + \frac{1}{2}kr^2.$$

Solving $V(r) = E$ gives $r_{\min}$ and $r_{\max}$ which are semi-axes.

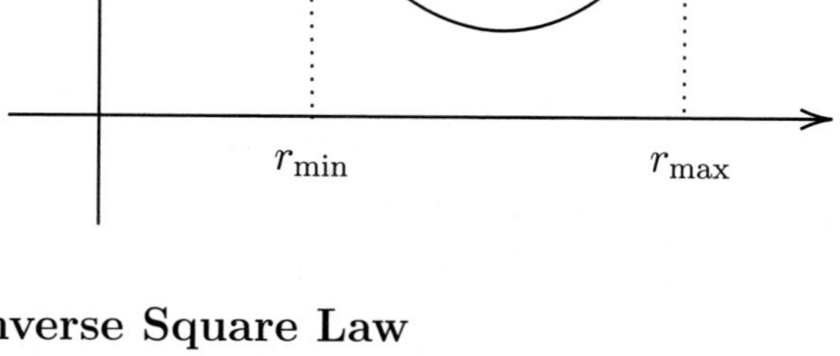

1.4.2 The Inverse Square Law

Now we consider the Kepler's Problem with potential

$$U(r) = -\frac{k}{r}$$

and the force is

$$\vec{F} = -\frac{k}{r^2}\frac{\vec{r}}{r}. \qquad \text{(inverse square growth)}$$

- If $k > 0$, the force is attractive, e.g., gravitational force and attractive electrostatic force.
- If $k < 0$, the force is repulsive, e.g., repulsive electrostatic force.

Let us discuss the attractive force. Here we assume $k > 0$. For simplicity, assume the motion is confined on the xy-plane

$$\vec{r} = (x_1, x_2, x_3 = 0) = (r\cos\theta, r\sin\theta, 0).$$

From the general discussion in Section 1.2.4, we have

$$\begin{cases} \dfrac{d\theta}{dt} = \dfrac{J}{mr^2} & J : \text{angular momentum} \\ \dfrac{dr}{dt} = \sqrt{\dfrac{2(E - V(r))}{m}} & E : \text{total energy} \end{cases}$$

where

$$V(r) = \frac{J^2}{2mr^2} - \frac{k}{r}$$

is the effective potential. So

$$\begin{aligned} \frac{d\theta}{dr} &= \frac{J/r^2}{\sqrt{2m(E - V(r))}} = \frac{J/r^2}{\sqrt{2mE - J^2/r^2 + \frac{2mk}{r}}} \\ &= \frac{J/r^2}{\sqrt{-\left(\frac{J}{r} - \frac{mk}{J}\right)^2 + 2mE + \frac{m^2k^2}{J^2}}} \end{aligned}$$

$$\Longrightarrow \qquad d\theta = \frac{-d\left(\frac{J}{r} - \frac{mk}{J}\right)}{\sqrt{-\left(\frac{J}{r} - \frac{mk}{J}\right)^2 + 2mE + \frac{m^2k^2}{J^2}}}.$$

Using

$$\int \frac{dx}{\sqrt{a^2 - x^2}} = -\arccos\left(\frac{x}{a}\right) + \text{const},$$

we can solve

$$\theta = \arccos\left(\frac{\frac{J}{r} - \frac{mk}{J}}{\sqrt{2mE + \frac{m^2k^2}{J^2}}}\right) + \theta_0.$$

We can reparametrize θ by shifting the origin of θ, such that $\theta_0 = 0$. Then

$$J/r = \frac{mk}{J} + \sqrt{2mE + \frac{m^2k^2}{J^2}}\cos\theta, \qquad \frac{J^2/mk}{r} = 1 + \sqrt{1 + \frac{2EJ^2}{mk^2}}\cos\theta.$$

Let $l = J^2/mk$, and $e = \sqrt{1 + \frac{2EJ^2}{mk^2}}$, then we get

$$\frac{l}{r} = 1 + e\cos\theta$$

or

$$r = \frac{l}{1 + e\cos\theta}$$

Here l is the latus rectum, and e is the eccentricity. Let us rewrite the equation in

$$\begin{cases} x = r\cos\theta \\ y = r\sin\theta \end{cases}$$

The equation becomes

$$\begin{aligned} r = l - er\cos\theta, \qquad & x^2 + y^2 = (l - ex)^2, \\ \Longrightarrow (1 - e^2)x^2 + 2lex + y^2 = l^2 & \\ \Longrightarrow (1 - e^2)(x + \frac{le}{1 - e^2})^2 + y^2 = \frac{l^2}{1 - e^2}. & \end{aligned}$$

There are three cases corresponding to

$$e < 1\,(E < 0), \qquad e = 1\,(E = 0), \qquad e > 1\,(E > 0).$$

This can be also distinguished from the effective potential

$$V(r) = \frac{J^2}{2mr^2} - \frac{k}{r}.$$

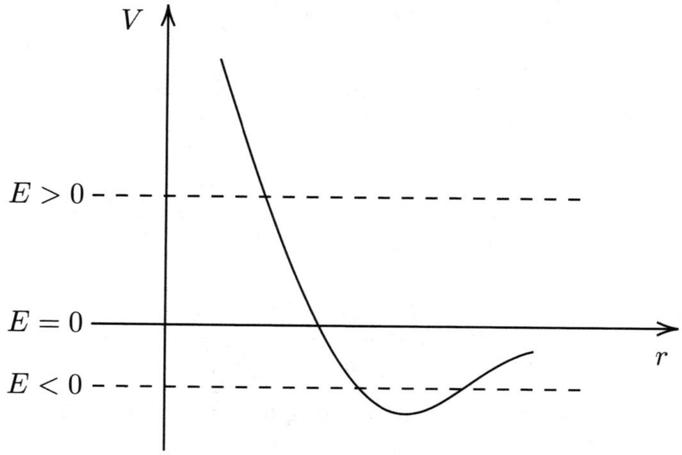

- $E < 0$, $e < 1$.

  The orbit is an ellipse

$$\left(x + \frac{le}{1 - e^2}\right)^2 + \frac{y^2}{1 - e^2} = \frac{l^2}{(1 - e^2)^2}.$$

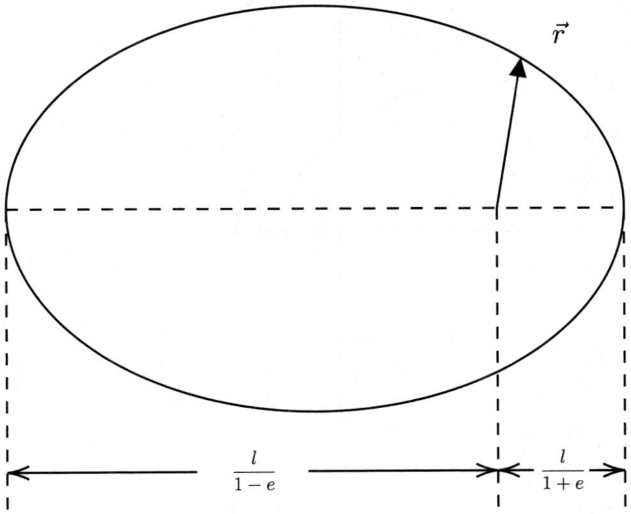

- $E = 0$, $e = 1$.

  The orbit is a parabola

$$2lx = l^2 - y^2.$$

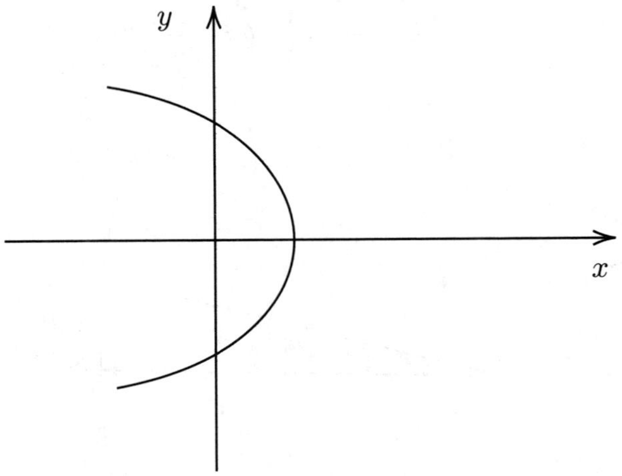

- $E > 0$, $e > 1$.

  The orbit is a hyperbola

$$\left(x - \frac{le}{e^2 - 1}\right)^2 - \frac{y^2}{e^2 - 1} = \frac{l^2}{(e^2 - 1)^2}.$$

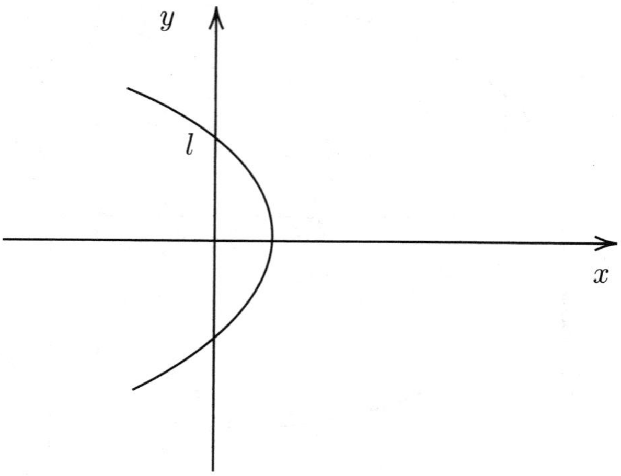

We have discovered Kepler's first law.

There is another interesting property about the angular momentum conservation. The rate of sweeping out area is

$$dA = \frac{1}{2} r \cdot r d\theta.$$

So

$$\frac{dA}{dt} = \frac{1}{2} r^2 \dot{\theta} = \frac{J}{2m}$$

is a constant.

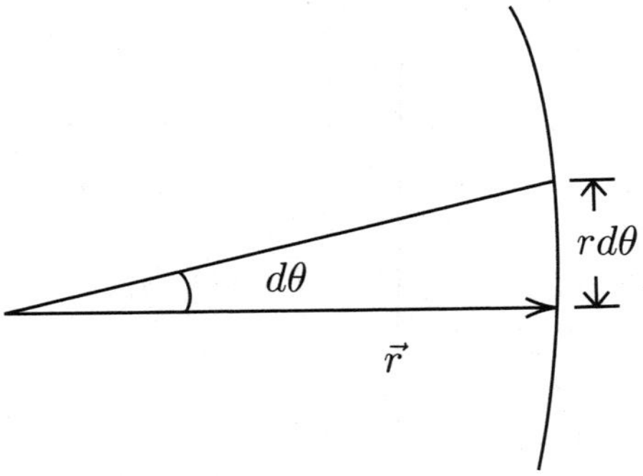

This is Kepler's second law.

Consider the elliptic orbit case ($E < 0$, $e < 1$). The total area is πab where a, b are semi-axes. Let τ denote the period. Since $\frac{dA}{dt} = \frac{J}{2m}$,

$$\pi ab = \frac{J}{2m} \tau \qquad \Longrightarrow \qquad \tau = \frac{2m\pi ab}{J}.$$

Now from the equation

$$\left(x + \frac{le}{1-e^2}\right)^2 + \frac{y^2}{1-e^2} = \frac{l^2}{(1-e^2)^2},$$

we have

$$a = \frac{l}{1-e^2}, \qquad b = \frac{l}{\sqrt{1-e^2}} = \sqrt{la}.$$

Since

$$e = \sqrt{1 + \frac{2EJ^2}{mk^2}}, \qquad l = J^2/mk, \qquad 1 - e^2 = \frac{2|E|J^2}{mk^2},$$

$$\Longrightarrow \qquad a = \frac{l}{1-e^2} = \frac{k}{2|E|}, \qquad b^2 = la = \frac{J^2}{2m|E|}.$$

So

$$\left(\frac{\tau}{2\pi}\right)^2 = \frac{m^2a^2b^2}{J^2} = \frac{m^2a^3l}{mkl} = \frac{ma^3}{k},$$

i.e.,

$$\left(\frac{\tau}{2\pi}\right)^2 = \frac{ma^3}{k}.$$

For gravitational force, we have $k = GMm$ where M is the mass of the center. It follows that

$$\left(\frac{\tau}{2\pi}\right)^2 = \frac{a^3}{GM}.$$

This is Kepler's third law.

1.4.3 Laplace-Runge-Lenz Vector

For a particle moving in $\mathbb{R}^3$ under central conservative force, we have conservation of energy and angular momentum. However, when the force goes like $\frac{1}{r^2}$, there is an extra conserved quantity. We explore this hidden symmetry for inverse square law. Let

$$\vec{r}(t) = (x_1(t), x_2(t), x_3(t)) \in \mathbb{R}^3, \qquad r = \sqrt{x_1^2 + x_2^2 + x_3^2}$$

denote the position vector. Let $V(r) = -\frac{k}{r}$ be the potential, where k is a constant. Let

$$\vec{J} = m\vec{r} \times \dot{\vec{r}}$$

be the angular momentum which is conserved. Define **Laplace-Runge-Lenz Vector**

$$\vec{A} = \frac{\dot{\vec{r}} \times \vec{J}}{k} - \frac{\vec{r}}{r}.$$

Using $\vec{a} \times \left(\vec{b} \times \vec{c}\right) = (\vec{a} \cdot \vec{c})\,\vec{b} - \left(\vec{a} \cdot \vec{b}\right)\vec{c}$ and the motion equation $m\ddot{\vec{r}} = -\frac{k\vec{r}}{r^3}$, we calculate

$$\begin{aligned}
\frac{d\vec{A}}{dt} &= \frac{1}{k}\left(\ddot{\vec{r}} \times \left(m\vec{r} \times \dot{\vec{r}}\right) + \dot{\vec{r}} \times \left(m\vec{r} \times \ddot{\vec{r}}\right)\right) - \frac{d}{dt}\left(\frac{\vec{r}}{r}\right) \\
&= \frac{1}{k}\left(\left(\ddot{\vec{r}} \cdot \dot{\vec{r}}\right) m\vec{r} - \left(\ddot{\vec{r}} \cdot m\vec{r}\right)\dot{\vec{r}} + \left(\dot{\vec{r}} \cdot \ddot{\vec{r}}\right) m\vec{r} - \left(\dot{\vec{r}} \cdot m\vec{r}\right)\ddot{\vec{r}}\right) - \frac{d}{dt}\left(\frac{\vec{r}}{r}\right) \\
&= -\frac{\vec{r} \cdot \dot{\vec{r}}}{r^3}\vec{r} + \frac{1}{r}\dot{\vec{r}} - \left(\dot{\vec{r}} \cdot \frac{\vec{r}}{r^3}\right)\vec{r} + \left(\dot{\vec{r}} \cdot \vec{r}\right)\frac{\vec{r}}{r^3} - \frac{d}{dt}\left(\frac{\vec{r}}{r}\right) \\
&= 0.
\end{aligned}$$

So the Laplace-Runge-Lenz Vector is conserved!

Using $\left(\vec{a} \times \vec{b}\right) \cdot \vec{c} = (\vec{c} \times \vec{a}) \cdot \vec{b}$, we have

$$\vec{A} \cdot \vec{r} = \frac{1}{k}\left(\dot{\vec{r}} \times \vec{J}\right) \cdot \vec{r} - r = \frac{1}{k}\left(\vec{r} \times \dot{\vec{r}}\right) \cdot \vec{J} - r = \frac{\vec{J}^2}{km} - r.$$

Let $\vec{A} \cdot \vec{r} = Ar\cos\theta$, where A is the length of $\vec{A}$ and θ is the angle between $\vec{A}$ and $\vec{r}$. Then

$$r = \left(\frac{\vec{J}^2}{km}\right)\frac{1}{1 + A\cos\theta}.$$

This also proves that the orbit in the Kepler problem must be an ellipse, parabola or hyperbola.

The conserved Laplace-Runge-Lenz Vector indicates a hidden symmetry from the perspective of Noether's Theorem. In fact, let $\vec{\epsilon} = (\epsilon_1, \epsilon_2, \epsilon_3)$ be three infinitesimal numbers. Consider the current system with Lagrangian $\mathscr{L} = \frac{1}{2}m\dot{\vec{r}}^2 + \frac{k}{r}$ and the infinitesimal variation

$$\delta_{\vec{\epsilon}}\vec{r} = \frac{1}{2}\vec{\epsilon} \times \left(\dot{\vec{r}} \times \vec{r}\right) + \frac{1}{2}\left(\vec{\epsilon} \times \dot{\vec{r}}\right) \times \vec{r}.$$

Here we have organized three infinitesimal variations into a single expression, and shall derive three conserved quantities all together. Let us compute the variation of the Lagrangian. Firstly

$$\delta_{\vec{\epsilon}}\dot{\vec{r}} = \frac{1}{2}\vec{\epsilon} \times \left(\ddot{\vec{r}} \times \vec{r}\right) + \frac{1}{2}\left(\vec{\epsilon} \times \ddot{\vec{r}}\right) \times \vec{r} + \frac{1}{2}\left(\vec{\epsilon} \times \dot{\vec{r}}\right) \times \dot{\vec{r}}.$$

The variation of the Kinetic term is

$$\begin{aligned}
\delta_{\vec{\epsilon}}\left(\frac{1}{2}m\dot{\vec{r}}^2\right) &= \frac{m}{2}\dot{\vec{r}} \cdot \left[\vec{\epsilon} \times \left(\ddot{\vec{r}} \times \vec{r}\right) + \left(\vec{\epsilon} \times \ddot{\vec{r}}\right) \times \vec{r} + \left(\vec{\epsilon} \times \dot{\vec{r}}\right) \times \dot{\vec{r}}\right] \\
&= \frac{m}{2}\left[\vec{\epsilon} \cdot \left(\left(\ddot{\vec{r}} \times \vec{r}\right) \times \dot{\vec{r}}\right) + \left(\vec{\epsilon} \times \ddot{\vec{r}}\right) \cdot \left(\vec{r} \times \dot{\vec{r}}\right)\right] \\
&= \frac{m}{2}\vec{\epsilon} \cdot \left[\left(\ddot{\vec{r}} \times \vec{r}\right) \times \dot{\vec{r}} + \ddot{\vec{r}} \times \left(\vec{r} \times \dot{\vec{r}}\right)\right] \\
&= \frac{m}{2}\vec{\epsilon} \cdot \frac{d}{dt}\left(\left(\dot{\vec{r}} \times \vec{r}\right) \times \dot{\vec{r}}\right).
\end{aligned}$$

The variation of the potential term is

$$\begin{aligned}
\delta_{\vec{\epsilon}}\left(\frac{k}{r}\right) &= -\frac{k}{r^2}\frac{\vec{r}\cdot\delta_{\vec{\epsilon}}\vec{r}}{r} = -\frac{k}{2r^3}\vec{r}\cdot\left[\vec{\epsilon}\times\left(\dot{\vec{r}}\times\vec{r}\right)+\left(\vec{\epsilon}\times\dot{\vec{r}}\right)\times\vec{r}\right] \\
&= -\frac{k}{2r^3}\vec{\epsilon}\cdot\left(\left(\dot{\vec{r}}\times\vec{r}\right)\times\vec{r}\right) = \frac{k}{2r^3}\vec{\epsilon}\cdot\left[r^2\dot{\vec{r}}-\left(\vec{r}\cdot\dot{\vec{r}}\right)\vec{r}\right] \\
&= \frac{k}{2}\vec{\epsilon}\left[\frac{\dot{\vec{r}}}{r}-\frac{\vec{r}}{r^3}\vec{r}\cdot\dot{\vec{r}}\right] \\
&= \frac{k}{2}\vec{\epsilon}\cdot\frac{d}{dt}\left(\frac{\vec{r}}{r}\right).
\end{aligned}$$

It follows that

$$\begin{aligned}
\delta_{\vec{\epsilon}}\mathscr{L} &= \frac{m}{2}\vec{\epsilon}\cdot\frac{d}{dt}\left(\left(\dot{\vec{r}}\times\vec{r}\right)\times\dot{\vec{r}}\right)+\frac{k}{2}\vec{\epsilon}\cdot\frac{d}{dt}\left(\frac{\vec{r}}{r}\right) \\
&= \vec{\epsilon}\cdot\frac{d}{dt}\left(\frac{\dot{\vec{r}}\times\vec{J}}{2}+\frac{k}{2}\frac{\vec{r}}{r}\right).
\end{aligned}$$

So the above variations are infinitesimal symmetries. Let us apply Theorem 1.3.4 to compute the conserved quantities. Since

$$\begin{aligned}
\frac{\partial\mathscr{L}}{\partial\dot{q}}\delta_{\vec{\epsilon}}\vec{r} &= \frac{m}{2}\dot{\vec{r}}\cdot\left[\vec{\epsilon}\times\left(\dot{\vec{r}}\times\vec{r}\right)+\left(\vec{\epsilon}\times\dot{\vec{r}}\right)\times\vec{r}\right] \\
&= \frac{m}{2}\left[\vec{\epsilon}\cdot\left(\left(\dot{\vec{r}}\times\vec{r}\right)\times\dot{\vec{r}}\right)+\left(\vec{\epsilon}\times\dot{\vec{r}}\right)\cdot\left(\vec{r}\times\dot{\vec{r}}\right)\right] \\
&= \frac{m}{2}\vec{\epsilon}\cdot\left[\left(\dot{\vec{r}}\times\vec{r}\right)\times\dot{\vec{r}}+\dot{\vec{r}}\times\left(\vec{r}\times\dot{\vec{r}}\right)\right] \\
&= m\vec{\epsilon}\cdot\left[\left(\dot{\vec{r}}\times\vec{r}\right)\times\dot{\vec{r}}\right] = \vec{\epsilon}\cdot\left(\dot{\vec{r}}\times\vec{J}\right).
\end{aligned}$$

Thus we obtain conserved quantities from Noether's Theorem (Theorem 1.3.4)

$$\dot{\vec{r}}\times\vec{J}-\left(\frac{\dot{\vec{r}}\times\vec{J}}{2}+\frac{k}{2}\frac{\vec{r}}{r}\right)=\frac{k}{2}\vec{A}$$

which are precisely related to the Laplace-Runge-Lenz vector.

Remark. Conservation of angular momentum is due to rotation in $\mathbb{R}^3$: symmetry under $SO(3)$. But the inverse square force law actually has bigger symmetry $SO(4)$, whose conserved quantities are angular momentum and the Laplace-Runge-Lenz vector.

1.5 Rigid Body

We consider extended objects with no internal degrees of freedom, called **rigid bodies**. A rigid body is described by a collection of points such that the distance between points is fixed:

$$|\vec{r}_i-\vec{r}_j| = \text{constant}$$

for any two points i, j.

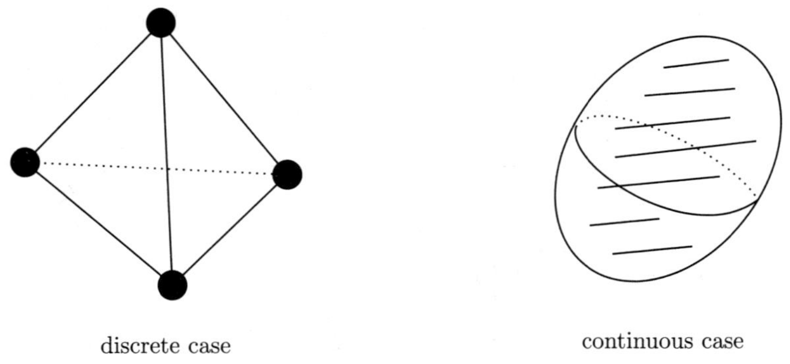

The motion of a rigid body has 6 degrees of freedom

$$3 \text{ translations} + 3 \text{ rotations}$$

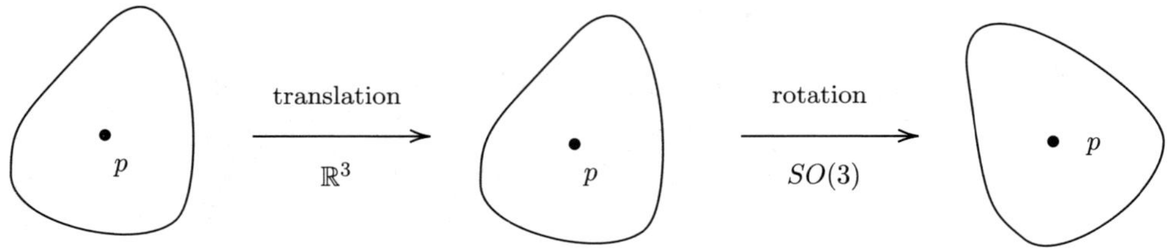

1.5.1 Angular Velocity

Let us first consider the body X with a point $p \in X$ fixed, and so X rotates about p. Let us fix a frame $\{\hat{e}_i\}$ in the space and choose a moving frame $\{e_i(t)\}$ in the body that moves with the body.

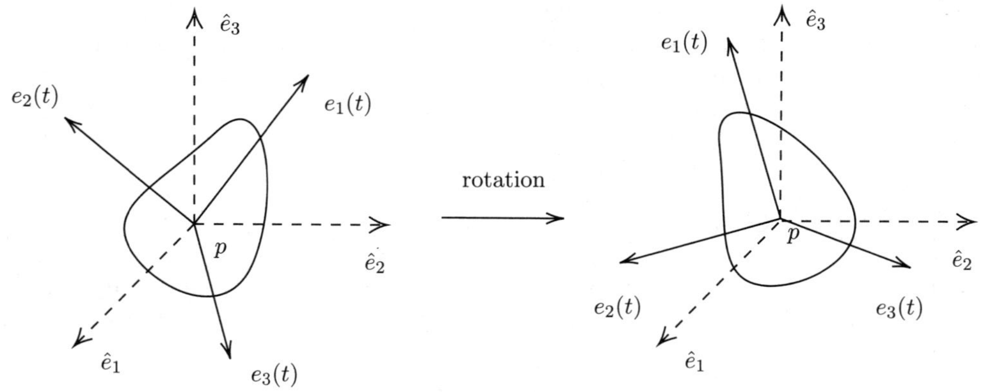

Both frames are orthonormal:

$$\hat{e}_i \cdot \hat{e}_j = \delta_{ij}, \qquad e_i \cdot e_j = \delta_{ij}.$$

We can express $\{e_i(t)\}$ in terms of $\{\hat{e}_i\}$ by a matrix

$$e_i(t) = \sum_{i,j} R_{ij}(t)\hat{e}_j, \qquad \text{or in matrix form,} \qquad \begin{pmatrix} e_1(t) \\ e_2(t) \\ e_3(t) \end{pmatrix} = R(t) \begin{pmatrix} \hat{e}_1 \\ \hat{e}_2 \\ \hat{e}_3 \end{pmatrix}.$$

$R(t)$ is an element of 3×3 special orthogonal matrices

$$SO(3) = \{A : 3 \times 3 \text{ matrix} \,|\, A^T A = I, \det A = 1\}.$$

In fact, using

$$\delta_{ij} = e_i(t) \cdot e_j(t) = \sum_{k,l} (R_{ik}\hat{e}_k) \cdot (R_{jl}\hat{e}_l) = \sum_{k,l} R_{ik}R_{jl}\delta_{kl} = \sum_k R_{ik}R_{jk},$$

we get $R^T R = I$. On the other hand,

$$1 = \det(R^T R) = \det(R)\det(R^T) = \det(R)^2,$$

So $\det(R) = \pm 1$. Since det is continuous, and det$= 1$ when $\{e_i\}$ coincides with $\{\hat{e}_i\}$, we have

$$\det(R) = 1.$$

Now any point $\vec{\gamma}$ in the body can be described by

$$\vec{\gamma}(t) = \sum_i \gamma_i e_i(t), \qquad \gamma_1, \gamma_2, \gamma_3 \text{ fixed},$$

so

$$\frac{d\vec{\gamma}}{dt} = \sum_i \gamma_i \frac{de_i(t)}{dt}.$$

We have

$$\frac{de_i(t)}{dt} = \sum_k \frac{d}{dt}(R_{ik}(t))\hat{e}_k = \sum_{k,j} \frac{d}{dt}(R_{ik}(t))(R^{-1})_{kj}(t)e_j(t) = \sum_j \omega_{ij}(t)e_j(t)$$

where

$$\omega_{ij} = \sum_k \left(\frac{d}{dt}R_{ik}\right)(R^{-1})_{kj} = \sum_k \left(\frac{d}{dt}R_{ik}\right) R_{jk}.$$

Here we have used $R^{-1} = R^T$. In terms of matrix,

$$\omega = \left(\frac{d}{dt}R\right) R^{-1} = \left(\frac{d}{dt}R\right) R^T.$$

We next show that ω is an element of the Lie algebra $so(3)$ of $SO(3)$:

$$so(3) = \{A : 3 \times 3 \text{ matrix} \,|\, A^T = -A\}$$

Proposition 1.5.1. *$\omega \in so(3)$, i.e., $\omega^T = -\omega$.*

Proof.

$$\omega^T = \left(\left(\frac{d}{dt}R\right)R^T\right)^T = R\frac{d}{dt}(R^T) = \frac{d}{dt}(R \cdot R^T) - \left(\frac{d}{dt}R\right)R^T$$
$$= -\left(\frac{d}{dt}R\right)R^T = -\omega. \qquad \square$$

Since ω is anti-symmetric, we can express

$$\omega_{ij} = \sum_k \epsilon_{ijk}\Omega_k \qquad \text{or} \qquad \Omega_i = \sum_{j,k} \frac{1}{2}\epsilon_{ijk}\omega_{jk}.$$

Here ϵ_{ijk} is the Levi-Civita symbol with $\epsilon_{123} = 1$. The vector $\{\Omega_i\}$ and the matrix ω collect the same amount of information. In fact, we have explicitly

$$\begin{cases} \Omega_1 = \omega_{23} = -\omega_{32} \\ \Omega_2 = \omega_{31} = -\omega_{13} \\ \Omega_3 = \omega_{12} = -\omega_{21} \end{cases}$$

In terms of Ω_i, and using $e_i \times e_j = \sum_k \epsilon_{ijk}e_k$,

$$\frac{de_i}{dt} = \sum_j \omega_{ij}e_j = \sum_{jk} \Omega_k\epsilon_{ijk}e_j = \sum_{jk} \Omega_k\epsilon_{kij}e_j = \sum_k \Omega_k e_k \times e_i = \vec{\Omega} \times e_i.$$

Here we define the vector

$$\vec{\Omega} = \sum_i \Omega_k e_k.$$

Now for any point $\vec{r}(t) = \sum_i r_i e_i(t)$ in the body,

$$\frac{d}{dt}\vec{r}(t) = \sum_i r_i \frac{d}{dt}e_i(t) = \sum_i r_i\vec{\Omega} \times e_i = \Omega \times \vec{r}(t),$$

we get

$$\frac{d\vec{r}}{dt} = \vec{\Omega} \times \vec{r}$$

The vector $\vec{\Omega}$ is called the **angular velocity**.

1.5.2 Inertia Tensor

We consider the kinetic energy for a rotation body around a fixed point p as above.

$$K = \frac{1}{2}\sum_\alpha m_\alpha \dot{\vec{r}}_\alpha^2 = \frac{1}{2}\sum_\alpha m_\alpha \left(\vec{\Omega} \times \vec{r}_\alpha\right) \cdot \left(\vec{\Omega} \times \vec{r}_\alpha\right)$$
$$= \frac{1}{2}\sum_\alpha m_\alpha \left(\left(\vec{\Omega} \cdot \vec{\Omega}\right)(\vec{r}_\alpha \cdot \vec{r}_\alpha) - \left(\vec{\Omega} \cdot \vec{r}_\alpha\right)^2\right) = \frac{1}{2}\sum_i \Omega_i I_{ij}\Omega_j,$$

where

$$I_{ij} = \sum_{\alpha} m_\alpha \left((\vec{r}_\alpha \cdot \vec{r}_\alpha)\, \delta_{ij} - (\vec{r}_\alpha)_i \, (\vec{r}_\alpha)_j \right).$$

This is called the **inertia tensor** of the body frame. It is a symmetric tensor: $I_{ij} = I_{ji}$. Moreover, they are independent of time. We can write I_{ij} in a matrix form as

$$I = \sum_{\alpha} m_\alpha \begin{pmatrix} y_\alpha^2 + z_\alpha^2 & -x_\alpha y_\alpha & -x_\alpha z_\alpha \\ -x_\alpha y_\alpha & x_\alpha^2 + z_\alpha^2 & -y_\alpha z_\alpha \\ -x_\alpha z_\alpha & -y_\alpha z_\alpha & x_\alpha^2 + y_\alpha^2 \end{pmatrix},$$

where $\vec{r}_\alpha = (x_\alpha, y_\alpha, z_\alpha)$.

In the case of continuous body, we can replace

$$\sum_{\alpha} \longrightarrow \int$$

and get the analogous expression

$$I = \int dxdydz\, \rho(x,y,z) \begin{pmatrix} y^2 + z^2 & -xy & -xz \\ -xy & x^2 + z^2 & -yz \\ -xz & -yz & x^2 + y^2 \end{pmatrix},$$

where ρ is the mass density of the body.

Proposition 1.5.2. *I is real, symmetric and semi-positive.*

Proof. Exercise. □

Observe that if $\{e_i'\}$ is a new moving body frame, it is related to $\{e_i\}$ by a constant matrix

$$e_i' = \sum_j \Theta_{ij} e_j$$

where $\Theta \in SO(3)$ is a special orthogonal matrix. Then the angular velocity $\vec{\Omega}'$ and the inertia tensor I' expressed in the new frame is related to $\vec{\Omega}$ and I by

$$I' = \Theta I \Theta^{-1} = \Theta I \Theta^T, \qquad \vec{\Omega}' = \Theta \vec{\Omega}.$$

Since I is a symmetric real semipositive matrix, we can always find a $\Theta \in SO(3)$ such that

$$\Theta I \Theta^{-1} = \begin{pmatrix} I_1 & & \\ & I_2 & \\ & & I_3 \end{pmatrix}$$

is diagonal. Equivalently, we can find a special body frame such that the inertia tensor becomes

$$I = \begin{pmatrix} I_1 & & \\ & I_2 & \\ & & I_3 \end{pmatrix}, \qquad I_i \geq 0.$$

The directions of such special frame are called the **principal axes of inertia**, and the corresponding diagonal components $\{I_i\}$ are called **principal moments of inertia**. In this frame,

$$K = \frac{1}{2}\sum_{i=1}^{3} I_i \Omega_i^2.$$

Example 1.5.3 (Disc). *We consider a uniform disc of radius R and mass M. We choose the body frame such that the disc is placed on the xy-plane centered at the origin. The mass density is*

$$\frac{M}{\pi R^2}\delta(z)dxdydz.$$

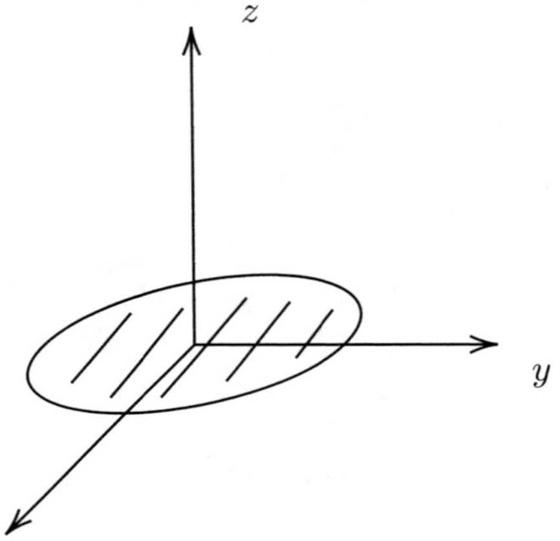

$$I = \int_D \frac{M}{\pi R^2} dxdy \begin{pmatrix} y^2 & -xy & 0 \\ -xy & x^2 & 0 \\ 0 & 0 & x^2+y^2 \end{pmatrix}$$

$$\overset{\text{By symmetry}}{=\!=\!=} \frac{M}{\pi R^2}\int_D dxdy \begin{pmatrix} y^2 & & \\ & x^2 & \\ & & x^2+y^2 \end{pmatrix} = \begin{pmatrix} I_1 & & \\ & I_2 & \\ & & I_3 \end{pmatrix},$$

$$I_1 = \frac{M}{\pi R^2}\int_{x^2+y^2\le R^2} dxdy\, y^2 = \frac{M}{2\pi R^2}\int_{x^2+y^2\le R^2} dxdy\,(x^2+y^2)$$

$$= \frac{M}{2\pi R^2}\int_0^{2\pi} d\theta \int_0^R rdr\, r^2 = \frac{1}{4}MR^2,$$

$$\Longrightarrow \qquad I_1 = I_2 = \frac{1}{4}MR^2, \qquad I_3 = I_1 + I_2 = \frac{1}{2}MR^2.$$

In general, the motion of a body is an overall translation superposed with a rotation. Let us describe the case when p is chosen to be the center of mass of the body. The motion of a point α of the body is described by

$$\vec{r}_\alpha(t) = \vec{R}(t) + \Delta\vec{r}_\alpha(t).$$

Here $\vec{R}(t)$ is the motion of the center of mass p, and $\triangle\vec{r}_\alpha$ is the position vector relative to p in the moving body.

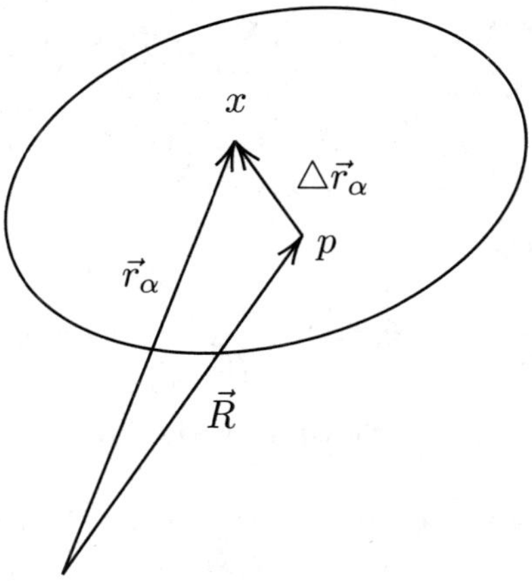

We have

$$\dot{\vec{R}}(t) = \frac{d}{dt}\vec{R}(t)$$

which is the velocity of center of mass, and

$$\triangle\dot{\vec{r}}_\alpha(t) = \frac{d}{dt}\triangle\vec{r}_\alpha(t) = \vec{\Omega} \times \triangle\vec{r}_\alpha$$

as before. The kinetic energy in general is

$$\begin{aligned} K = \frac{1}{2}\sum_\alpha m_\alpha \dot{\vec{r}}_\alpha^{\,2} &= \frac{1}{2}\sum_\alpha m_\alpha \left(\dot{\vec{R}} + \vec{\Omega} \times \triangle\vec{r}_\alpha\right) \cdot \left(\dot{\vec{R}} + \vec{\Omega} \times \triangle\vec{r}_\alpha\right) \\ &= \frac{1}{2}\sum_\alpha m_\alpha \dot{\vec{R}}^2 + \dot{\vec{R}} \cdot \left(\vec{\Omega} \times \sum_\alpha m_\alpha \triangle\vec{r}_\alpha\right) + \frac{1}{2}\vec{\Omega}^T I \vec{\Omega}. \end{aligned}$$

By the definition of center of mass,

$$\sum_\alpha m_\alpha \triangle\vec{r}_\alpha = 0.$$

So

$$K = \frac{1}{2}M\dot{\vec{R}}^2 + \frac{1}{2}\vec{\Omega}^T I \vec{\Omega}.$$

Here $M = \sum_\alpha m_\alpha$ is the total mass. We see that the dynamics separates into the motion of the center of mass $\vec{R}$, together with rotation about the center of mass. If we consider some point other than the center of mass, we can also obtain a formula which we do not discuss here.

1.5.3 Euler's Equation

We assume the body is rotating about a fixed point p. The angular momentum is

$$\begin{aligned}\vec{J} &= \sum_\alpha m_\alpha \vec{r}_\alpha \times \dot{\vec{r}}_\alpha = \sum_\alpha m_\alpha \vec{r}_\alpha \times \left(\vec{\Omega} \times \vec{r}_\alpha\right) \\ &= \sum_\alpha m_\alpha \left((\vec{r}_\alpha \cdot \vec{r}_\alpha)\vec{\Omega} - \left(\vec{\Omega} \cdot \vec{r}_\alpha\right)\vec{r}_\alpha\right) \\ &= I\vec{\Omega}.\end{aligned}$$

We find

$$\vec{J} = I\vec{\Omega}$$

Note that in general $\vec{J} \neq \vec{\Omega}$, i.e., the spin of the body is different from its angular momentum.

Now we assume the body is rotating freely. Then the Lagrangian is

$$K = \frac{1}{2}\vec{\Omega}^T I \vec{\Omega}.$$

Since the body is free, the angular momentum is conserved:

$$\frac{d\vec{J}}{dt} = 0.$$

Assume we have chose the moving body frame $\{e_i\}$ to be the principal axes of inertia. Then

$$\vec{\Omega} = \sum_i \Omega_i e_i, \qquad I = \begin{pmatrix} I_1 & & \\ & I_2 & \\ & & I_3 \end{pmatrix}.$$

Let us also expand $\vec{J}$ in terms of $\{e_i\}$ by

$$\vec{J} = \sum_i J_i e_i.$$

Angular momentum conservation gives

$$0 = \frac{d\vec{J}}{dt} = \sum_i \left(\frac{d}{dt}J_i\right) e_i + \sum_i J_i \Omega \times e_i$$

$$\Longrightarrow \qquad \frac{d}{dt}J_i + \sum_{j,k} \epsilon_{ijk}\Omega_j J_k = 0.$$

On the other hand, $\vec{J} = I\vec{\Omega}$ leads to

$$J_i = I_i \Omega_i.$$

Plugging into the above, we find

$$I_i\dot{\Omega}_i + \sum_{j,k} \epsilon_{ijk}\Omega_j\Omega_k I_k = 0.$$

Equivalently,

$$\begin{cases} I_1\dot{\Omega}_1 + \Omega_2\Omega_3(I_3 - I_2) = 0 \\ I_2\dot{\Omega}_2 + \Omega_3\Omega_1(I_1 - I_3) = 0 \\ I_3\dot{\Omega}_3 + \Omega_1\Omega_2(I_2 - I_1) = 0 \end{cases}$$

These are called **Euler's equations**.

1.5.4 Free Tops

We next analyze the motion of free rotating bodies (also called **free tops**) using Euler's equation.

a). Sphere. In this case, we have $I_1 = I_2 = I_3$, so Ω is constant. The sphere spins around the same axis.

b). Symmetric top. In this case, $I_1 = I_2 \neq I_3$. So we have

$$\begin{cases} I_1\dot{\Omega}_1 = \Omega_2\Omega_3(I_1 - I_3) \\ I_1\dot{\Omega}_2 = -\Omega_1\Omega_3(I_1 - I_3) \\ I_3\dot{\Omega}_3 = 0 \end{cases}$$

We find Ω_3 is a constant of motion. Let $\alpha = \Omega_3(I_1 - I_3)/I_1$. Then

$$\begin{cases} \dot{\Omega}_1 = \alpha\Omega_2 \\ \dot{\Omega}_2 = -\alpha\Omega_1 \end{cases}$$

The solution is

$$(\Omega_1, \Omega_2) = (A \sin\alpha t,\ A \cos\alpha t).$$

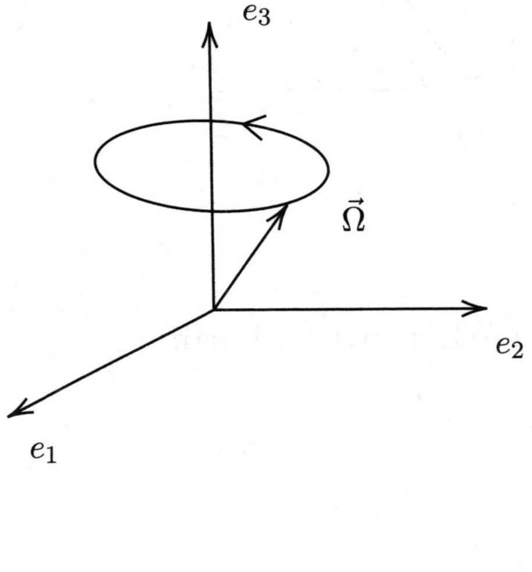

Here e_3 is the direction of the symmetric axis. The direction of the spin processes about the e_3-axis with frequency α.

c). Asymmetric top. In this case, $I_1 < I_2 < I_3$.

The general case is more complex. We consider the conservation of energy and angular momentum

$$\begin{cases} I_1\Omega_1^2 + I_2\Omega_2^2 + I_3\Omega_3^2 = 2K \\ I_1^2\Omega_1^2 + I_2^2\Omega_2^2 + I_3^2\Omega_3^2 = J^2 \end{cases}$$

Here J is the magnitude of the angular momentum. We can also rewrite it in terms of $J_i = I_i\Omega_i$:

$$\begin{cases} \dfrac{J_1^2}{2KI_1} + \dfrac{J_2^2}{2KI_2} + \dfrac{J_3^2}{2KI_3} = 1 \\ J_1^2 + J_2^2 + J_3^2 = J^2 \end{cases}$$

The first is an ellipsoid with major axis J_3 and minor axis J_1. The second is a sphere. When $\vec{J} = \sum_i J_i e_i$ is expressed relative to the axes of inertia, it moves on the interaction of these two surfaces. For fixed K and J^2, the body will spin in a stable manner around the principal axes with the smallest and largest moments of inertia, but not around the intermediate axis.

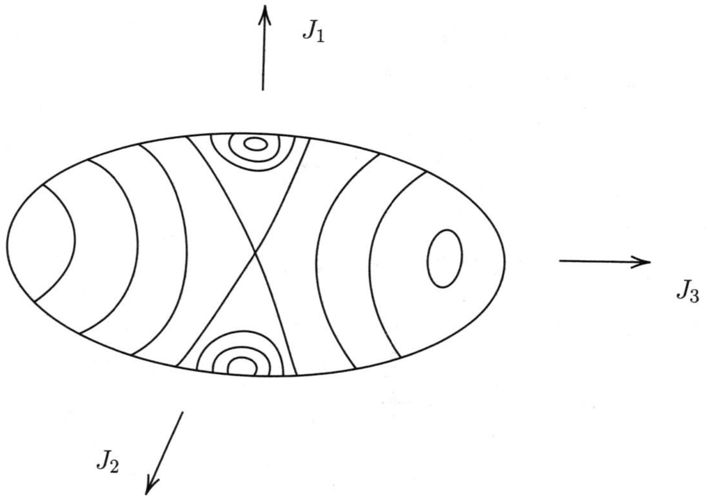

1.5.5 Euler's Equation in Lax Form

Recall

$$\vec{J} = I\vec{\Omega}$$

which is conserved in the free case. Let us rewrite Euler's equation in terms of the matrix $\omega = (\omega_{ij})$ where

$$\omega_{ij} = \sum_k \epsilon_{ijk}\Omega_k, \qquad \text{or} \qquad \omega = \begin{pmatrix} 0 & \Omega_3 & -\Omega_2 \\ -\Omega_3 & 0 & \Omega_1 \\ \Omega_2 & -\Omega_1 & 0 \end{pmatrix}.$$

We also introduce the matrix $\varphi = (\varphi_{ij})$

$$\varphi_{ij} = \sum_k \epsilon_{ijk}J_k, \qquad \text{or} \qquad \varphi = \begin{pmatrix} 0 & J_3 & -J_2 \\ -J_3 & 0 & J_1 \\ J_2 & -J_1 & 0 \end{pmatrix}.$$

Euler's equation reads

$$\frac{dJ_i}{dt} = \sum_{j,k} \epsilon_{ijk}J_j\Omega_k = \sum_j \omega_{ij}J_j$$

$$\Longrightarrow \qquad \frac{d\varphi_{ij}}{dt} = \sum_k \epsilon_{ijk}\sum_m \omega_{km}J_m = \frac{1}{2}\sum_{k,m,l,s} \epsilon_{ijk}\omega_{km}\epsilon_{mls}\varphi_{ls}.$$

In the above summation, the term

$$\epsilon_{ijk}\omega_{km}\epsilon_{mls}$$

needs $k \neq m$ for nonzero contribution, so

$$m = i \qquad \text{or} \qquad m = j.$$

Therefore

$$\begin{aligned}
\frac{d\varphi_{ij}}{dt} &= \frac{1}{2}\sum_{k,l,s} \epsilon_{ijk}\omega_{ki}\epsilon_{ils}\varphi_{ls} + \frac{1}{2}\sum_{k,l,s} \epsilon_{ijk}\omega_{kj}\epsilon_{jls}\varphi_{ls} \\
&= \frac{1}{2}\sum_{k,l,s} \omega_{ki}(\delta_{jl}\delta_{ks} - \delta_{js}\delta_{kl})\varphi_{ls} + \frac{1}{2}\sum_{k,l,s} \omega_{kj}(-\delta_{il}\delta_{ks} + \delta_{is}\delta_{kl})\varphi_{ls} \\
&= \frac{1}{2}\sum_k \omega_{ki}(\varphi_{jk} - \varphi_{kj}) + \frac{1}{2}\sum_k \omega_{kj}(-\varphi_{ik} + \varphi_{ki}) \\
&= \sum_k \omega_{ik}\varphi_{kj} - \sum_k \varphi_{ik}\omega_{kj}.
\end{aligned}$$

In matrix form, this is

$$\frac{d\varphi}{dt} = [\omega, \varphi].$$

Here $[\omega, \varphi] = \omega\varphi - \varphi\omega$ is the commutator.

This is a version of so-called **Lax equation**, which is a fundamental equation in integrable systems. As one illustration, we immediately have

$$\operatorname{Tr}\varphi^n$$

as a constant of motion for any n. In fact, using the trace property $\mathrm{Tr}(AB) = \mathrm{Tr}(BA)$, we have

$$\frac{d}{dt}\mathrm{Tr}\,\varphi^n = n\mathrm{Tr}\,\dot{\varphi}\varphi^{n-1} = n\mathrm{Tr}\,[\omega,\varphi]\varphi^{n-1} = \mathrm{Tr}\,[\omega,\varphi^n] = 0.$$

For example,

$$\mathrm{Tr}\,\varphi = 0, \qquad \mathrm{Tr}\,\varphi^2 = -2(J_1^2 + J_2^2 + J_3^2).$$

However, the energy E can not be obtained from $\mathrm{Tr}\,\varphi^n$. But the Lax equation can be modified to suffice this as we show below.

We can also express the relation

$$\vec{J} = I\vec{\Omega}$$

in terms of φ and ω. In fact, let

$$\Lambda = \frac{1}{2}\begin{pmatrix} I_2 + I_3 - I_1 & & \\ & I_1 + I_3 - I_2 & \\ & & I_1 + I_2 - I_3 \end{pmatrix}.$$

Then

$$\varphi = \Lambda\omega + \omega\Lambda.$$

We have arrived at the following equation

$$\frac{d\varphi}{dt} = [\omega,\varphi], \qquad \varphi = \Lambda\omega + \omega\Lambda.$$

Here Λ is a constant diagonal matrix related to the principal inertia.

Now we show the above Lax equation can be modified via a spectral parameter. Introduce

$$L(t,z) = \Lambda^2 + \frac{\varphi(t)}{z}, \qquad A(t,z) = z\Lambda + \omega(t),$$

where z is a new variable. Direct calculation shows

$$\frac{\partial L}{\partial t} = [A, L]$$

for any z. Thus we have found a new form of Lax equation for rigid body. In particular, $\mathrm{Tr}\,L^n$ is a constant of motion for any n. Let us compute

$$\mathrm{Tr}\,L^2 = \mathrm{Tr}\,(\Lambda^2 + \varphi/z)^2 = \mathrm{Tr}\,\Lambda^4 + \frac{2}{z}\mathrm{Tr}\,\Lambda^2\varphi + \frac{1}{z^2}\mathrm{Tr}\,\varphi^2.$$

Since $\mathrm{Tr}\,L^2$ is conserved for every z, we get conservation of $\mathrm{Tr}\,\varphi^2$ again.

$$\mathrm{Tr}\,L^3 = \mathrm{Tr}\,(\Lambda^2 + \varphi/z)^3 = \mathrm{Tr}\,\Lambda^6 + \frac{3}{z}\mathrm{Tr}\,\Lambda^4\varphi + \frac{3}{z^2}\mathrm{Tr}\,\Lambda^2\varphi^2 + \frac{1}{z^3}\mathrm{Tr}\,\varphi^3.$$

So the coefficient of the order $\frac{1}{z^2}$

$$\mathrm{Tr}\,\Lambda^2\varphi^2 = 2EI_1I_2I_3 - \frac{1}{2}(J_1^2 + J_2^2 + J_3^2)(I_1^2 + I_2^2 + I_3^2)$$

is conserved. We get conservation of the energy $E = K$ as promised.

Chapter 2

Hamiltonian Mechanics

2.1 Hamilton's Equations

2.1.1 Hamilton's Equations

Recall that in Lagrangian formulation, we have the action functional

$$S = \int \mathscr{L}(q, \dot{q}, t)\, dt.$$

The extremal path of S leads to the equation of motion (Euler-Lagrange equation)

$$\frac{d}{dt}\left(\frac{\partial \mathscr{L}}{\partial \dot{q}_i}\right) - \frac{\partial \mathscr{L}}{\partial q_i} = 0.$$

Definition 2.1.1. The generalized momentum conjugate to q_i is defined to be

$$p_i = \frac{\partial \mathscr{L}}{\partial \dot{q}_i}, \qquad i = 1, \cdots, n.$$

Remark. We have seen that in Newton mechanics with

$$\mathscr{L} = \frac{1}{2} m \dot{q}^2 - V(q),$$

the momentum conjugate to q_i is

$$p_i = m \dot{q}_i$$

which is the usual mechanical momentum. In general, the conjugate momentum may have different forms, see Example 1.3.6 for charged particles in electro-magnetic backgrounds. Also, if the system has translation symmetry along q_i, then p_i is conserved.

In Lagrangian mechanics, we use

$$q_i, \dot{q}_i$$

as independent variables to describe the Lagrangian $\mathscr{L} = \mathscr{L}(q, \dot{q}, t)$. In Hamiltonian mechanics, we use

$$q_i, p_i$$

as independent variables to describe the system. They are related by the equation

$$p_i = \frac{\partial \mathscr{L}}{\partial \dot{q}_i},$$

from which we can solve to find the relation

$$\dot{q}_i = \dot{q}_i(q, p).$$

The pair (q_i, p_i) defines a point in a $2n$-dimensional space, called the "phase space". The advantage of Hamiltonian mechanics is that it has close relationship with geometry (symplectic geometry) and links the classical and quantum theory.

Definition 2.1.2. Define the Hamiltonian function of q, p and t by

$$\mathscr{H}(q, p, t) = \sum_{i=1}^{n} p_i \dot{q}_i(q, p) - \mathscr{L}(q, \dot{q}(q, p), t).$$

Let us consider the total variation of $\mathscr{H}$.

- On one side, we have

$$d\mathscr{H} = \sum_i \frac{\partial \mathscr{H}}{\partial q_i} dq_i + \sum_i \frac{\partial \mathscr{H}}{\partial p_i} dp_i + \frac{\partial \mathscr{H}}{\partial t} dt.$$

- On the other hand, using $p_i = \frac{\partial \mathscr{L}}{\partial \dot{q}_i}$, we have

$$\begin{aligned} d\mathscr{H} &= \sum_i (\dot{q}_i dp_i + p_i d\dot{q}_i) - \sum_i \left(\frac{\partial \mathscr{L}}{\partial q_i} dq_i + \frac{\partial \mathscr{L}}{\partial \dot{q}_i} d\dot{q}_i + \frac{\partial \mathscr{L}}{\partial t} dt \right) \\ &= -\sum_i \frac{\partial \mathscr{L}}{\partial q_i} dq_i + \sum_i \dot{q}_i dp_i - \frac{\partial \mathscr{L}}{\partial t} dt. \end{aligned}$$

Comparing the above two expressions, we conclude that

$$\begin{cases} \dfrac{\partial \mathscr{H}}{\partial q_i} = -\dfrac{\partial \mathscr{L}}{\partial q_i} \\ \dfrac{\partial \mathscr{H}}{\partial p_i} = \dot{q}_i \\ \dfrac{\partial \mathscr{H}}{\partial t} = -\dfrac{\partial \mathscr{L}}{\partial t} \end{cases}$$

Now the Euler-Lagrange equation

$$\frac{\partial \mathscr{L}}{\partial q_i} = \frac{d}{dt}\left(\frac{\partial \mathscr{L}}{\partial \dot{q}_i} \right) = \frac{d}{dt} p_i$$

is equivalent to the following system of equations

$$\begin{cases} \dot{q}_i = \dfrac{\partial \mathscr{H}}{\partial p_i} \\ \dot{p}_i = -\dfrac{\partial \mathscr{H}}{\partial q_i} \end{cases}$$

These are **Hamilton's equations.** We have replaced n second order differential equations (Euler-Lagrange) by $2n$ first order differential equations (Hamilton). At any time, the "state" of the system is defined by a point (q_i, p_i) in the phase space. Hamilton equation allows us to determine the state at all time in the future from an initial state.

Another way to see the equivalence between Euler-Lagrange equation and Hamilton equation is through the principle of least action directly. Let

$$r(t) = (q(t), p(t))$$

be any path in the phase space. Define the action as a functional of path $(q(t), p(t))$ by

$$S[q(t), p(t)] = \int_{t_0}^{t_1} \left(\sum_i p_i \dot{q}_i - \mathscr{H} \right) dt = \int_{t_0}^{t_1} \left(\sum_i p_i dq_i - \mathscr{H} dt \right).$$

Consider the variation δS with fixed endpoints

$$\begin{aligned} \delta S &= \int_{t_0}^{t_1} \sum_i \delta p_i dq_i + p_i d(\delta q_i) - \delta \mathscr{H}\, dt \\ &\xlongequal[\text{by part}]{\text{integration}} \int_{t_0}^{t_1} \sum_i \delta p_i dq_i - \delta q_i dp_i - \left(\frac{\partial \mathscr{H}}{\partial p_i} \delta p_i + \frac{\partial \mathscr{H}}{\partial q_i} \delta q_i \right) dt \\ &= \int_{t_0}^{t_1} \left(\sum_i \left(\dot{q}_i - \frac{\partial \mathscr{H}}{\partial p_i} \right) \delta p_i - \sum_i \left(\dot{p}_i + \frac{\partial \mathscr{H}}{\partial q_i} \right) \delta q_i \right) dt. \end{aligned}$$

Requiring $\delta S = 0$ for arbitrary variations, we find

$$\begin{cases} \dot{q}_i = \dfrac{\partial \mathscr{H}}{\partial p_i} \\ \dot{p}_i = -\dfrac{\partial \mathscr{H}}{\partial q_i} \end{cases}$$

which is precisely the Hamilton's equations.

Assume $(q_i(t), p_i(t))$ satisfies Hamilton's equation, then

$$\begin{aligned} \frac{d}{dt} \mathscr{H}(q(t), p(t), t) &= \sum_i \left(\frac{\partial \mathscr{H}}{\partial q_i} \dot{q}_i + \frac{\partial \mathscr{H}}{\partial p_i} \dot{p}_i \right) + \frac{\partial \mathscr{H}}{\partial t} \\ &= \sum_i \left(\frac{\partial \mathscr{H}}{\partial q_i} \frac{\partial \mathscr{H}}{\partial p_i} - \frac{\partial \mathscr{H}}{\partial p_i} \frac{\partial \mathscr{H}}{\partial q_i} \right) + \frac{\partial \mathscr{H}}{\partial t} \\ &= \frac{\partial \mathscr{H}}{\partial t}. \end{aligned}$$

Thus if $\mathscr{H}$ does not depend on t explicitly, i.e., $\frac{\partial \mathscr{H}}{\partial t} = 0$, then $\frac{d\mathscr{H}}{dt} = 0$ and so $\mathscr{H}$ is conserved under Hamilton's equations. This is the law of conservation of energy.

In fact, in Newton mechanics,

$$\mathscr{L} = \frac{1}{2} m\dot{q}^2 - V(q), \qquad p_i = \frac{\partial \mathscr{L}}{\partial \dot{q}_i} = m\dot{q}_i,$$

the Hamiltonian function

$$\mathscr{H} = \sum_i p_i \dot{q}_i - \mathscr{L} = m\dot{q}^2 - \mathscr{L} = \frac{1}{2} m\dot{q}^2 + V(q)$$

is precisely the total energy. We have just rederived the energy conservation via Hamilton's equations.

Example 2.1.3 (Particles in conservative force).

$$\mathscr{L} = \frac{1}{2} m\dot{q}^2 - V(q), \qquad p_i = m\dot{q}_i, \qquad \mathscr{H} = \frac{1}{2} m\dot{q}^2 + V(q) = \frac{p^2}{2m} + V(q).$$

Hamilton's equations read

$$\begin{cases} \dot{q}_i = \dfrac{\partial \mathscr{H}}{\partial p_i} = \dfrac{1}{m} p_i \\ \dot{p}_i = -\dfrac{\partial \mathscr{H}}{\partial q_i} = -\dfrac{\partial V}{\partial q_i} \end{cases}$$

The first equation defines the momentum. The second equation describes Newton's law.

Example 2.1.4 (Charged particle in electromagnetic field).

$$\mathscr{L} = \frac{1}{2} m\dot{\vec{x}}^2 - Q\left(\phi - \vec{A}\cdot\dot{\vec{x}}\right), \qquad \vec{x} = (x_1, x_2, x_3).$$

The momentum conjugate to x_i is

$$p_i = \frac{\partial \mathscr{L}}{\partial \dot{x}_i} = m\dot{x}_i + QA_i.$$

This allows us to solve

$$\dot{x}_i = \frac{1}{m}(p_i - QA_i).$$

The Hamiltonian function is

$$\begin{aligned} \mathscr{H}(x,p) &= \sum_i p_i \dot{x}_i - \mathscr{L} \\ &= \vec{p}\cdot\frac{1}{m}\left(\vec{p} - Q\vec{A}\right) - \left(\frac{1}{2m}\left(\vec{p} - Q\vec{A}\right)^2 - Q\phi + \frac{Q}{m}\vec{A}\cdot\left(\vec{p} - Q\vec{A}\right)\right) \\ &= \frac{\left(\vec{p} - Q\vec{A}\right)^2}{2m} + Q\phi. \end{aligned}$$

Hamilton's equations read

$$\begin{cases} \dot{x}_i = \dfrac{\partial \mathscr{H}}{\partial p_i} = \dfrac{1}{m}(p_i - QA_i) \\ \dot{p}_i = -\dfrac{\partial \mathscr{H}}{\partial x_i} = \dfrac{Q}{m}\left(\vec{p} - Q\vec{A}\right) \cdot \partial_i \vec{A} - Q\partial_i \phi \end{cases}$$

2.1.2 Legendre Transform

The relationship between Lagrangian and Hamiltonian formalism can be described by the Legendre transform. Let $f : \mathbb{R}^n \to \mathbb{R}$ be a smooth function. Assume f is **convex**, so the matrix $\frac{\partial^2 f}{\partial x_i \partial x_j}$ is positive definite everywhere. The Legendre transform of f is a function f^* on $p \in (\mathbb{R}^n)^* = \mathbb{R}^n$ defined in the following way:

$$f^*(p) = \sup_{x \in \mathbb{R}^n} \{p \cdot x - f(x)\}.$$

Since f is convex, there is a unique x such that the sup is achieved (when it exists). Let us call it $x(p)$. Geometrically, this is the largest vertical distance between the line $y = px$ and the graph of $y = f(x)$.

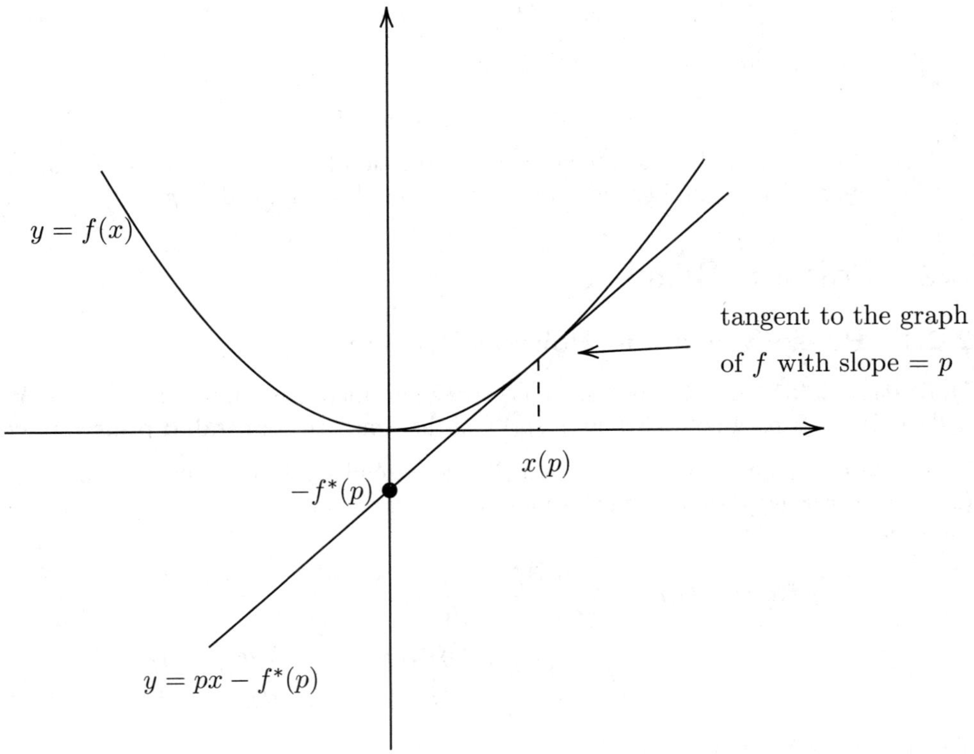

The point $x(p)$ is solved by

$$\frac{\partial}{\partial x}(p \cdot x - f(x)) = 0,$$

or equivalently

$$p = \nabla f.$$

Solving the equation of x in terms of p, we get

$$x = x(p).$$

Note that

$$f^*(p) = px(p) - f(x(p)) \qquad \Longrightarrow \qquad \frac{\partial f^*}{\partial p_i} = x_i + p \cdot \frac{\partial}{\partial p_i} x(p) - \nabla f \cdot \frac{\partial x}{\partial p_i} = x_i.$$

We find the symmetric relation:

$$\begin{cases} p = \nabla f \\ x = \nabla f^* \end{cases}$$

In fact, we have a duality theorem

Theorem 2.1.5. *Let f be a convex function above. Then its Legendre transformation f^* is also a convex function, whose Legendre transformation gives back f, i.e.,*

$$(f^*)^* = f.$$

Now the Hamiltonian

$$\mathscr{H}(q, p, t) = p\dot{q} - \mathscr{L}(q, \dot{q}, t)$$

with $p = \frac{\partial \mathscr{L}}{\partial \dot{q}}$ is precisely the Legendre transformation from $\dot{q}$ to p. In particular, $\mathscr{L}$ can be obtained from the Legendre transformation of $\mathscr{H}$ applied to p.

2.2 Poisson Bracket

2.2.1 Phase Space and Poisson Bracket

Definition 2.2.1. Let $\mathbb{R}^{2n}$ be the phase space parameterized by $\{q_i, p_i\}_{i=1,\cdots,n}$. We will call $\mathbb{R}^{2n} \times \mathbb{R}$ parameterized by (q_i, p_i) and the time t the **extended phase space**.

Let $f(q, p, t)$ be a smooth function on the extended phase space. Assume now $(q(t), p(t))$ satisfies Hamilton's equations. Consider the evolution $f(q(t), p(t), t)$ along the time:

$$\begin{aligned} \frac{d}{dt} f(q(t), p(t), t) &= \sum_i \left(\frac{\partial f}{\partial q_i} \dot{q}_i + \frac{\partial f}{\partial p_i} \dot{p}_i \right) + \frac{\partial f}{\partial t} \\ &\overset{\text{Hamilton}}{\underset{\text{equation}}{=\!=\!=}} \sum_i \left(\frac{\partial f}{\partial q_i} \frac{\partial \mathscr{H}}{\partial p_i} - \frac{\partial f}{\partial p_i} \frac{\partial \mathscr{H}}{\partial q_i} \right) + \frac{\partial f}{\partial t}. \end{aligned}$$

This motivates the following

Definition 2.2.2. Given two functions f, g on the (extended) phase space, we define their **Poisson bracket** by

$$\{f, g\} = \sum_i \left(\frac{\partial f}{\partial q_i} \frac{\partial g}{\partial p_i} - \frac{\partial f}{\partial p_i} \frac{\partial g}{\partial q_i} \right).$$

From the above computation, we find

Proposition 2.2.3. *If $(q(t), p(t))$ satisfies Hamilton's equation, then*

$$\frac{d}{dt} f(q(t), p(t), t) = \{f, \mathscr{H}\} + \frac{\partial f}{\partial t}.$$

In particular, $f(q(t), p(t), t)$ is a conserved quantity if and only if

$$\frac{\partial f}{\partial t} + \{f, \mathscr{H}\} = 0.$$

Now we study several properties of the Poisson bracket $\{-, -\}$.

Proposition 2.2.4. *The Poisson bracket satisfies the following properties:*

- *Antisymmetry:* $\{f, g\} = -\{g, f\}$.
- *Bilinear:* $\{f, \lambda_1 g + \lambda_2 h\} = \lambda_1 \{f, g\} + \lambda_2 \{f, h\}$, *for* $\lambda_i \in \mathbb{R}$.
- *Leibniz rule:* $\{f, gh\} = \{f, g\} h + g \{f, h\}$.
- *Jacobi identity:* $\{\{f, g\}, h\} + \{\{g, h\}, f\} + \{\{h, f\}, g\} = 0$.

Proof of Jacobi identity. Define the first order differential operator

$$D_f = \sum_i \left(\frac{\partial f}{\partial q_i} \frac{\partial}{\partial p_i} - \frac{\partial f}{\partial p_i} \frac{\partial}{\partial q_i} \right).$$

Then

$$\{f, g\} = D_f(g).$$

Observe that given two differential operators

$$A = \sum_i A_i(x) \frac{\partial}{\partial x^i}, \qquad B = \sum_i B_i(x) \frac{\partial}{\partial x^i},$$

their commutator as operators is computed by

$$[A, B] = AB - BA = \sum_i A(B_i(x)) \frac{\partial}{\partial x^i} - B(A_i(x)) \frac{\partial}{\partial x^i}.$$

Applying this to D_f, D_g, we find

$$\begin{aligned}
[D_f, D_g] &= D_f D_g - D_g D_f \\
&= \sum_i \left(D_f \left(\frac{\partial g}{\partial q_i} \right) \frac{\partial}{\partial p_i} - D_f \left(\frac{\partial g}{\partial p_i} \right) \frac{\partial}{\partial q_i} \right) \\
&\quad - \sum_i \left(D_g \left(\frac{\partial f}{\partial q_i} \right) \frac{\partial}{\partial p_i} - D_g \left(\frac{\partial f}{\partial p_i} \right) \frac{\partial}{\partial q_i} \right) \\
&= \sum_i \left(\left\{ f, \frac{\partial g}{\partial q_i} \right\} - \left\{ g, \frac{\partial f}{\partial q_i} \right\} \right) \frac{\partial}{\partial p_i} - \sum_i \left(\left\{ f, \frac{\partial g}{\partial p_i} \right\} - \left\{ g, \frac{\partial f}{\partial p_i} \right\} \right) \frac{\partial}{\partial q_i} \\
&= \sum_i \left(\left\{ \frac{\partial f}{\partial q_i}, g \right\} + \left\{ f, \frac{\partial g}{\partial q_i} \right\} \right) \frac{\partial}{\partial p_i} - \sum_i \left(\left\{ \frac{\partial f}{\partial p_i}, g \right\} + \left\{ f, \frac{\partial g}{\partial p_i} \right\} \right) \frac{\partial}{\partial q_i} \\
&= \sum_i \left(\frac{\partial}{\partial q_i} \{f, g\} \right) \frac{\partial}{\partial p_i} - \sum_i \left(\frac{\partial}{\partial p_i} \{f, g\} \right) \frac{\partial}{\partial q_i} \\
&= D_{\{f,g\}}.
\end{aligned}$$

Here we have used the simple fact that

$$\frac{\partial}{\partial q_i}\{f, g\} = \left\{ \frac{\partial f}{\partial q_i}, g \right\} + \left\{ f, \frac{\partial g}{\partial q_i} \right\}$$

and similarly for $\frac{\partial}{\partial p_i}$. So we have found

$$[D_f, D_g] = D_{\{f,g\}}.$$

Applying this to h, we have

$$D_{\{f,g\}} h = D_f D_g h - D_g D_f h, \qquad \text{i.e.,} \qquad \{\{f, g\}, h\} = \{f, \{g, h\}\} - \{g, \{f, h\}\},$$

$$\Longrightarrow \qquad \{\{f, g\}, h\} + \{\{g, h\}, f\} + \{\{h, f\}, g\} = 0.$$ □

Remark. The antisymmetry of $\{-, -\}$ implies that

$$\{f, f\} = 0, \qquad \forall f.$$

In particular, if we apply this to the Hamiltonian $\mathscr{H}$, then for $(q(t), p(t))$ satisfying Hamilton's equation, we have

$$\frac{d}{dt}\mathscr{H}(q(t), p(t), t) = \{\mathscr{H}, \mathscr{H}\} + \frac{\partial \mathscr{H}}{\partial t} = \frac{\partial \mathscr{H}}{\partial t}.$$

This gives another derivation of the evolution law of the Hamiltonian function.

Remark. We have the simple relations

$$\{q_i, q_j\} = \{p_i, p_j\} = 0, \qquad \{q_i, p_j\} = \delta_{ij} = -\{p_i, q_j\}.$$

In fact, the above relations together with the Leibniz rule determine the Poisson bracket $\{-, -\}$.

2.2.2 Constant of Motion

Definition 2.2.5. A function $f(q,p,t)$ on the extended phase space is called a **constant of motion** (or **integral of motion**) if

$$\frac{\partial}{\partial t}f + \{f, \mathscr{H}\} = 0.$$

Equivalently, for $(q(t), p(t))$ satisfying Hamilton's equation, we have

$$\frac{d}{dt}f(q(t), p(t), t) = 0.$$

In particular, if $f = f(q,p)$, then f is a constant of motion if and only if $\{f, \mathscr{H}\} = 0$. We say f and $\mathscr{H}$ "Poisson commute".

Proposition 2.2.6. *Let $f = f(q,p)$ and $g = g(q,p)$ be two functions on the phase space. Assume both f and g are constants of motion, then so is $\{f, g\}$.*

Proof.

$$\{\{f,g\}, \mathscr{H}\} = \{f, \{g, \mathscr{H}\}\} - \{g, \{f, \mathscr{H}\}\} = 0.$$

□

Example 2.2.7. *Consider a motion in $\mathbb{R}^3$ with central conservative force. The Lagrangian is*

$$\mathscr{L} = \frac{1}{2}m\dot{\vec{r}}^2 - U(r), \qquad \vec{r} = (x_1, x_2, x_3).$$

The conjugate momentums are

$$p_i = \frac{\partial \mathscr{L}}{\partial \dot{x}_i} = m\dot{x}_i.$$

The Hamiltonian is

$$\mathscr{H}(x,p) = \frac{\vec{p}^2}{2m} + U(r).$$

We have the angular momentum

$$\vec{J} = m\vec{r} \times \dot{\vec{r}} = \vec{r} \times \vec{p}.$$

In terms of components,

$$\begin{cases} J_1 = x_2p_3 - x_3p_2 \\ J_2 = x_3p_1 - x_1p_3 \\ J_3 = x_1p_2 - x_2p_1 \end{cases}$$

Let us check they are constants of motion in the phase space:

$$\begin{aligned}
\{J_1, \mathscr{H}\} &= \{x_2p_3 - x_3p_2, \mathscr{H}\} \\
&= x_2\{p_3, \mathscr{H}\} + p_3\{x_2, \mathscr{H}\} - x_3\{p_2, \mathscr{H}\} - p_2\{x_3, \mathscr{H}\} \\
&= -x_2\frac{\partial}{\partial x_3}\mathscr{H} + p_3\frac{\partial}{\partial p_2}\mathscr{H} + x_3\frac{\partial}{\partial x_2}\mathscr{H} - p_2\frac{\partial}{\partial p_3}\mathscr{H}
\end{aligned}$$

$$= \left(-x_2\frac{\partial r}{\partial x_3} + x_3\frac{\partial r}{\partial x_2}\right)U'(r) + \frac{1}{m}(p_3p_2 - p_2p_3)$$

$$\overset{r=\sqrt{x_1^2+x_2^2+x_3^2}}{=\!=\!=\!=\!=\!=} \left(\frac{-x_2x_3}{r} + \frac{x_3x_2}{r}\right)U'(r) = 0.$$

Similarly we can compute

$$\{J_2, \mathscr{H}\} = \{J_3, \mathscr{H}\} = 0.$$

So J_i's are constants of motion. On the other hand,

$$\{J_1, J_2\} = \{x_2p_3 - x_3p_2, x_3p_1 - x_1p_3\} = -x_2p_1 + x_1p_2 = J_3.$$

Similarly, we find the following Poisson bracket relations

$$\begin{cases} \{J_1, J_2\} = J_3 \\ \{J_2, J_3\} = J_1 \\ \{J_3, J_1\} = J_2 \end{cases}$$

So in fact once we know any two of J_i's are constants of motion, so is the third one!

Example 2.2.8. *Consider the 2-dim harmonic oscillator*

$$\mathscr{L} = \frac{1}{2}m(\dot{x}_1^2 + \dot{x}_2^2) - \frac{1}{2}k(x_1^2 + x_2^2).$$

The conjugate momentums are

$$p_1 = m\dot{x}_1, \qquad p_2 = m\dot{x}_2.$$

The Hamiltonian is

$$\mathscr{H} = \frac{1}{2m}(p_1^2 + m^2\omega^2x_1^2) + \frac{1}{2m}(p_2^2 + m^2\omega^2x_2^2) =: \mathscr{H}_1 + \mathscr{H}_2,$$

where $\omega = \sqrt{k/m}$. The angular momentum

$$J = x_1p_2 - x_2p_1$$

is a constant of motion as before. Let us consider the following quantities

$$\alpha_{ij} = \frac{1}{2m}(p_ip_j + m^2\omega^2x_ix_j).$$

We have

$$\alpha_{11} = \mathscr{H}_1, \qquad \alpha_{22} = \mathscr{H}_2.$$

It is obvious that

$$\{\alpha_{11}, \mathscr{H}\} = \{\mathscr{H}_1, \mathscr{H}_1\} = 0, \qquad \{\alpha_{22}, \mathscr{H}\} = \{\mathscr{H}_2, \mathscr{H}_2\} = 0.$$

For the off-diagonal element

$$\alpha_{12} = \alpha_{21} = \frac{1}{2m}(p_1 p_2 + m^2\omega^2 x_1 x_2),$$

we can compute directly

$$\{\alpha_{12}, \mathscr{H}\} = 0.$$

So $\{\alpha_{ij}\}$ *are all constants of motion. The solutions for the motion can be found*

$$\begin{cases} x_1 = \sqrt{\dfrac{2\alpha_{11}}{m\omega^2}}\sin(\omega t + \theta_1) \\ x_2 = \sqrt{\dfrac{2\alpha_{22}}{m\omega^2}}\sin(\omega t + \theta_2) \\ \alpha_{12} = \sqrt{\alpha_{11}\alpha_{22}}\cos(\theta_1 - \theta_2) \end{cases}$$

2.3 Liouville's Theorem

2.3.1 Phase Flow and Liouville's Theorem

We consider a Hamiltonian $\mathscr{H}$ on the phase space parameterized by (q_i, p_i). For simplicity, we assume $\mathscr{H}$ does not depend explicitly on the time t:

$$\mathscr{H} = \mathscr{H}(q, p).$$

Definition 2.3.1. We define the **phase flow** as the map on the phase space defined by

$$\phi_t : (q(0), p(0)) \longmapsto (q(t), p(t))$$

where $q(t)$, $p(t)$ are solutions of Hamilton's equations.

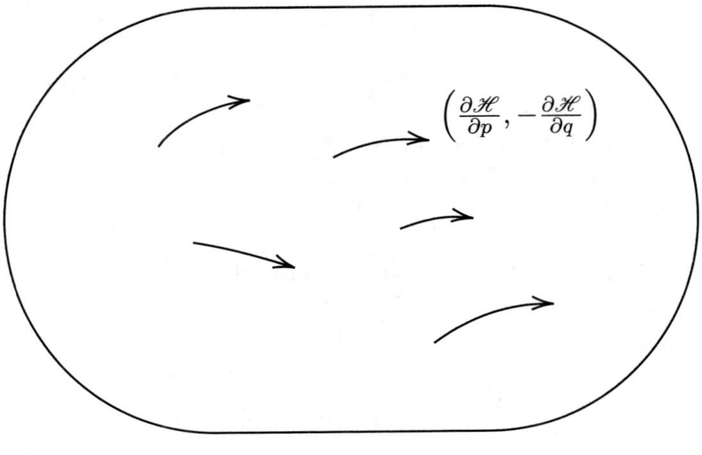

In other word, ϕ_t sends a point (q, p) in the phase space to a point that is obtained by the evolution of Hamilton's equations after time t with (q, p) being the initial point.

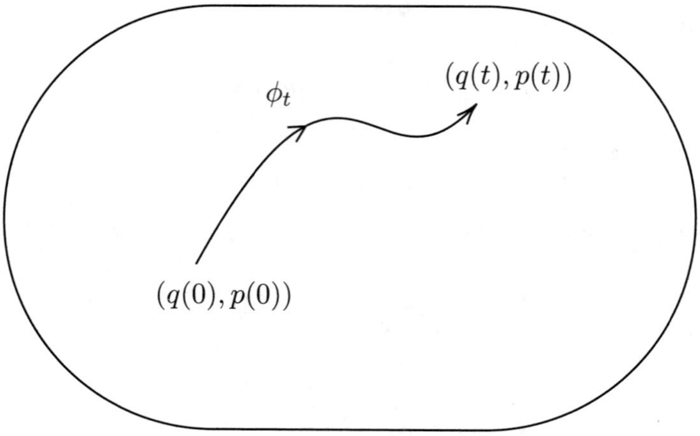

Proposition 2.3.2. *If $\mathscr{H} = \mathscr{H}(q, p)$ does not depend explicitly on t, then the phase flow has the following group property*

$$\phi_{t_1} \circ \phi_{t_2} = \phi_{t_1+t_2}.$$

This follows from the uniqueness of the solution of Hamilton's equations.

Theorem 2.3.3 (Liouville). *The phase flow preserves volume: for any region D, we have*

$$\text{volume}(\phi_t(D)) = \text{volume}(D).$$

Proof. Let $V(t) = \text{volume}(\phi_t(D))$. It is suffice to show that

$$\left.\frac{d}{dt}\right|_{t=0} V(t) = 0 \tag{†}$$

for any D. In fact, assume the equation (†) above. Then for any t_0,

$$\begin{aligned} \left.\frac{d}{dt}\right|_{t=t_0} V(t) &= \left.\frac{d}{dt}\right|_{t=0} V(t+t_0) = \left.\frac{d}{dt}\right|_{t=0} \text{volume}(\phi_{t+t_0}(D)) \\ &= \left.\frac{d}{dt}\right|_{t=0} \text{volume}(\phi_t(\phi_{t_0}(D))) = 0, \end{aligned}$$

so $V(t)$ is a constant. Equation (†) follows from the next lemma that we will show below. □

Lemma 2.3.4. *Consider the flow*

$$\phi_t : (x_1(0), \cdots, x_n(0)) \longmapsto (x_1(t), \cdots, x_n(t))$$

by the solution of the system of differential equations

$$\frac{d}{dt} x_i(t) = f_i(x(t)), \qquad i = 1, \cdots, n.$$

Let D be a region in $\mathbb{R}^n$ parametrized by $\{x_i\}$. Then

$$\left.\frac{d}{dt}\right|_{t=0} \int_{\phi_t(D)} d^n x = \int_D \left(\sum_i \partial_i f_i \right) d^n x = \int_D \operatorname{div}(\vec{f})\, d^n x.$$

Here $d^n x = dx_i \cdots dx_n$ is the standard volume form, and $\vec{f} = (f_1, \cdots, f_n)$.

Proof. The change of coordinate $\vec{x} = \vec{x}(0) \to \vec{x}(t)$ gives

$$\int_{\phi_t(D)} d^n x = \int_D \det \left(\frac{\partial \vec{x}(t)}{\partial \vec{x}} \right) d^n x.$$

Here $\left(\frac{\partial \vec{x}(t)}{\partial \vec{x}} \right)$ is the matrix whose (ij)-component is $\frac{\partial \vec{x}_i(t)}{\partial x_j}$. From the above equation

$$\frac{d}{dt} x_i(t) = f_i(x(t)),$$

we have

$$x_i(t) = x_i + t f_i(x) + O(t^2), \qquad x_i = x_i(0).$$

So

$$\frac{\partial \vec{x}_i(t)}{\partial x_j} = \delta_{ij} + t \partial_j f_i + O(t^2).$$

In terms of matrix, this is

$$\frac{\partial \vec{x}(t)}{\partial \vec{x}} = 1 + t \frac{\partial \vec{f}}{\partial \vec{x}} + O(t^2) = 1 + tA + O(t^2).$$

Then

$$\det \left(\frac{\partial \vec{x}(t)}{\partial \vec{x}} \right) = \det(1 + tA) + O(t^2) = 1 + t \operatorname{Tr} A + O(t^2) = 1 + t \sum_i \partial_i f_i + O(t^2),$$

$$\Longrightarrow \qquad \left.\frac{d}{dt}\right|_{t=0} \int_D \det \left(\frac{\partial \vec{x}(t)}{\partial \vec{x}} \right) d^n x = \int_D \left(\sum_i \partial_i f_i \right) d^n x. \qquad \square$$

Now we apply this lemma to proof of the theorem,

$$V(t) = \int_{\phi_t(D)} d^n q d^n p,$$

where

$$d^n q = dq_1 \cdots , dq_n, \qquad d^n p = dp_1 \cdots dp_n.$$

ϕ_t is given by the flow

$$\begin{cases} \dot{q}_i = \dfrac{\partial \mathscr{H}}{\partial p_i} \\ \dot{p}_i = -\dfrac{\partial \mathscr{H}}{\partial q_i} \end{cases}$$

So

$$\left.\frac{d}{dt}\right|_{t=0} V(t) = \int_D \left(\sum_i \frac{\partial}{\partial q_i} \left(\frac{\partial \mathscr{H}}{\partial p_i} \right) - \frac{\partial}{\partial p_i} \left(\frac{\partial \mathscr{H}}{\partial q_i} \right) \right) d^n q d^n p = 0.$$

Remark. The above lemma illustrate the equivalence of the following two viewpoints

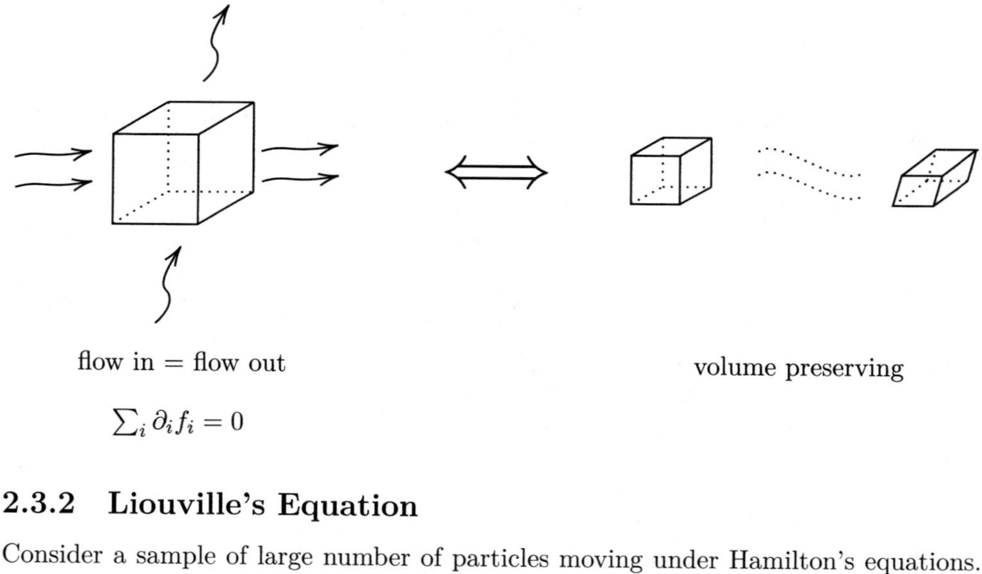

flow in = flow out

$\sum_i \partial_i f_i = 0$

volume preserving

2.3.2 Liouville's Equation

Consider a sample of large number of particles moving under Hamilton's equations. We plot them on the phase space and get the distribution function on the density of particles

$$\rho(q, p, t) d^n q d^n p.$$

Start with a small region D,

under the phase flow, the number of particles inside $\phi_t(D)$ is clearly fixed. On the other hand, the volume element d^nqd^np is conserved under the phase flow, so we conclude

$$\frac{d}{dt}\rho(q(t), p(t), t) = 0$$

under the phase flow. Therefore we have

$$\frac{\partial}{\partial t}\rho + \{\rho, \mathscr{H}\} = 0.$$

This is **Liouville's equation**, which is essentially another face of Liouville's Theorem. In particular, if ρ does not depend explicitly on t, we will have

$$\{\rho, \mathscr{H}\} = 0,$$

i.e., ρ is conserved. An important class of distribution is of the form

$$\rho = \rho(\mathscr{H}).$$

For example, the famous Boltzmann distribution has the form

$$\rho = \exp\left(-\frac{\mathscr{H}}{kT}\right),$$

where T is the temperature and k is the Boltzmann constant.

2.3.3 Poincaré's Recurrence Theorem

Theorem 2.3.5 (Poincaré's Recurrence Theorem). *Let ϕ be a volume-preserving continuous one-to-one mapping from a bounded region D to itself:*

$$\phi : D \longrightarrow D.$$

Then in any neighborhood U of any initial point of D, there is a point $x \in U$ which returns to U by ϕ, i.e., there exists $n > 0$ such that $\phi^n x \in U$.

Proof. Consider U, $\phi(U)$, $\phi^2(U)$, $\cdots$. Each has the same volume by assumption. Since $\mathrm{Vol}(U) < \infty$, there exist $k > l > 0$ such that

$$\phi^k(U) \cap \phi^l(U) \neq \varnothing \qquad \Longrightarrow \qquad \phi^{k-l}(U) \cap U \neq \varnothing.$$

Let $\phi^{k-l}(x) = y$ be one of the intersection point. Then such x gives the required point. □

This theorem in particular applies to the phase flow for mechanical system with bounded phase space (e.g. bounded by conserved energy).

Puzzle: Consider gas molecules all in one corner of the room. If we let them go, they will fill in the room. The theorem says if we wait long enough, they will almost all return once more to the corner of the room. The solution of the puzzle is that the waiting time is too long!

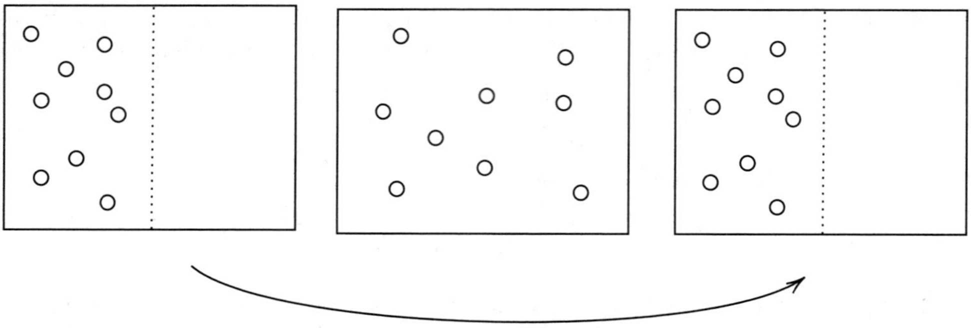

2.4 Canonical Transformation

Canonical transformation can be viewed as a coördinate transformation of the phase space that preserves the form of Hamilton's equations. We can use it to simplify the structure of Hamiltonian function and derive certain constants of motion.

To distinguish Poisson bracket under coordinate transformations, let us denote $\{-,-\}_{q,p}$ for the Poisson bracket defined by the phase coordinates $\{q_i, p_i\}$.

2.4.1 Time-independent Canonical Transformation

Let us start with time-independent canonical transformations where the Hamiltonian function remains the same. The time-dependent case will be discussed in Section 2.4.3 below.

Definition 2.4.1. A coordinate transformation on the phase space

$$\{q_i, p_i\} \longmapsto \{Q_i(q,p), P_i(q,p)\}$$

is called a **canonical transformation** if

$$\{Q_i, P_i\}_{p,q} = \delta_{ij}, \qquad \{Q_i, Q_j\}_{p,q} = \{P_i, P_j\}_{p,q} = 0.$$

In other words, the canonical transformation preserves Poisson brackets on coordinate generators. This property holds in fact for all functions. Let us denote

$$\{-,-\}_{p,q} : \text{Poisson bracket with respect to phase coordinates } q_i, p_i,$$

$$\{-,-\}_{P,Q} : \text{Poisson bracket with respect to phase coordinates } Q_i, P_i.$$

The coordinate transformation is

$$\{q_i, p_i\} \longmapsto \{Q_i(q,p), P_i(q,p)\}.$$

Given a function f on the phase space, we can write $f = f(Q,P)$ in terms of $\{Q_i, P_i\}$, or $f = f(Q(q,p), P(q,p))$ in terms of $\{q_i, p_i\}$. Then

$$\{f,g\}_{P,Q} = \sum_i \frac{\partial f}{\partial Q_i}\frac{\partial g}{\partial P_i} - \frac{\partial f}{\partial P_i}\frac{\partial g}{\partial Q_i}, \qquad \{f,g\}_{p,q} = \sum_i \frac{\partial f}{\partial q_i}\frac{\partial g}{\partial p_i} - \frac{\partial f}{\partial p_i}\frac{\partial g}{\partial q_i}.$$

Proposition 2.4.2. *$\{f,g\}_{P,Q} = \{f,g\}_{p,q}$ as functions on phase space if $\{q_i,p_i\} \mapsto \{Q_i,P_i\}$ is a canonical transformation. In other words, Poisson bracket is invariant under canonical transformations.*

Remark. Alternately, we can use $\{-,-\}_{p,q} = \{-,-\}_{P,Q}$ to define canonical transformations.

Proof. We prove the case when there is only one degree of freedom, i.e., the phase space has dimension 2. The proof in general is similar. Let us have a canonical transformation

$$\{q,p\} \longmapsto \{Q,P\}.$$

Then

$$\begin{aligned}
\{f,g\}_{p,q} &= \frac{\partial f}{\partial q}\frac{\partial g}{\partial p} - \frac{\partial f}{\partial p}\frac{\partial g}{\partial q} \\
&= \left(\frac{\partial P}{\partial q}\frac{\partial f}{\partial P} + \frac{\partial Q}{\partial q}\frac{\partial f}{\partial Q}\right)\left(\frac{\partial P}{\partial p}\frac{\partial g}{\partial P} + \frac{\partial Q}{\partial p}\frac{\partial g}{\partial Q}\right) \\
&\quad - \left(\frac{\partial P}{\partial p}\frac{\partial f}{\partial P} + \frac{\partial Q}{\partial p}\frac{\partial f}{\partial Q}\right)\left(\frac{\partial P}{\partial q}\frac{\partial g}{\partial P} + \frac{\partial Q}{\partial q}\frac{\partial g}{\partial Q}\right) \\
&= \left(\frac{\partial Q}{\partial q}\frac{\partial P}{\partial p} - \frac{\partial Q}{\partial p}\frac{\partial P}{\partial q}\right)\frac{\partial f}{\partial Q}\frac{\partial g}{\partial P} - \left(\frac{\partial Q}{\partial q}\frac{\partial P}{\partial p} - \frac{\partial Q}{\partial p}\frac{\partial P}{\partial q}\right)\frac{\partial f}{\partial P}\frac{\partial g}{\partial Q} \\
&= \{Q,P\}_{p,q}\frac{\partial f}{\partial Q}\frac{\partial g}{\partial P} - \{Q,P\}_{p,q}\frac{\partial f}{\partial P}\frac{\partial g}{\partial Q} \\
&= \frac{\partial f}{\partial Q}\frac{\partial g}{\partial P} - \frac{\partial f}{\partial P}\frac{\partial g}{\partial Q} = \{f,g\}_{P,Q}.
\end{aligned}$$

□

Corollary 2.4.3. *If $\{q_i,p_i\} \mapsto \{Q_i(q,p), P_i(q,p)\}$ is a canonical transformation, then its inverse map $\{Q_i,P_i\} \mapsto \{q_i(Q,P), p_i(Q,P)\}$ is also a canonical transformation. Moreover, a composition of canonical transformations is a canonical transformation. In particular, canonical transformations form a group.*

Proof. We prove for the inverse transformation. In fact

$$\{q_i,p_j\}_{P,Q} = \{q_i,p_j\}_{p,q} = \delta_{ij}, \qquad \{q_i,q_j\}_{P,Q} = \{q_i,q_j\}_{p,q} = 0 = \{p_i,p_j\}_{P,Q}.$$

□

In the original phase space coordinates (q_i,p_i), Hamilton's equation is

$$\frac{df}{dt} = \{f,\mathscr{H}\}_{p,q} + \frac{\partial f}{\partial t}.$$

We assume for simplicity $\mathscr{H} = \mathscr{H}(q,p)$. Let $\{q_i,p_i\} \mapsto \{Q_i,P_i\}$ be a canonical transformation. We can express $\mathscr{H}$ as a function of Q, P as

$$\mathscr{H} = \mathscr{H}(q(Q,P),p(Q,P)).$$

Then in the new coordinates $\{Q_i,P_i\}$ on the phase space,

$$\{f,\mathscr{H}\}_{P,Q} + \frac{\partial f}{\partial t}(q(Q,P),p(Q,P),t) = \{f,\mathscr{H}\}_{p,q} + \frac{\partial f}{\partial t}(q,p,t).$$

Therefore the evolution equation takes the same form in the Q_i, P_i coordinates:

$$\frac{df}{dt} = \{f, \mathscr{H}\}_{Q,P} + \frac{\partial f}{\partial t}.$$

In other words, the canonical transformation preserves the form of Hamilton's equation.

Also, from the invariance of Poisson bracket, if f is a constant of motion in (q_i, p_i) coordinates, then f is also a constant of motion in (Q_i, P_i) coordinates. This allows us to find conserved quantities by using a canonical transformation that simplifies the Hamiltonian function.

Example 2.4.4 (1-dim Harmonic Oscillator).

$$\mathscr{H} = \frac{p^2}{2m} + \frac{kq^2}{2} = \frac{1}{2m}(p^2 + m^2\omega^2 q^2), \qquad \omega = \sqrt{\frac{k}{m}}.$$

Consider a transformation of the form

$$q = \frac{f(P)}{m\omega}\sin Q, \qquad p = f(P)\cos Q,$$

where f is a function to be determined. This would lead to

$$\mathscr{H} = \frac{f^2(P)}{2m}$$

which is independent of Q. Now we check the condition for canonicity.

$$1 = \{q, p\}_{P,Q} = \frac{\partial q}{\partial Q}\frac{\partial p}{\partial P} - \frac{\partial q}{\partial P}\frac{\partial p}{\partial Q} = \frac{f(P)}{m\omega}\cos Q f'(P)\cos Q + \frac{f'(P)}{m\omega}\sin Q f(P)\sin Q$$
$$= \frac{f(P)f'(P)}{m\omega} = \frac{d}{dP}\left(\frac{f^2}{2m\omega}\right).$$

So we can solve f to find

$$f = \sqrt{2Pm\omega}.$$

Therefore

$$\begin{cases} q = \sqrt{\dfrac{2P}{m\omega}}\sin Q \\ p = \sqrt{2Pm\omega}\cos Q \end{cases}$$

is a canonical transformation. In (P, Q) coordinates, we have

$$\mathscr{H} = \omega P.$$

In particular, $P = \frac{\mathscr{H}}{\omega} = \frac{E}{\omega}$ is a constant of motion and

$$\frac{dQ}{dt} = \{Q, \mathscr{H}\} = \omega.$$

So $Q = \omega t + \varphi_0$ *under phase flow. The solution is*

$$\begin{cases} q = \sqrt{\dfrac{2P}{m\omega}} \sin(\omega t + \varphi_0) \\ p = \sqrt{2Pm\omega} \cos(\omega t + \varphi_0) \end{cases}$$

2.4.2 Infinitesimal Canonical Transformation

Consider transformations of the form

$$\begin{cases} q_i \longmapsto Q_i = q_i + \epsilon F_i(q,p) \\ p_i \longmapsto P_i = p_i + \epsilon E_i(q,p) \end{cases}$$

where ϵ is an infinitesimal small number. We ask when this generates an infinitesimal canonical transformation, i.e., a canonical transformation up to first order.

$$\begin{aligned} \{Q_i, P_j\} &= \{q_i + \epsilon F_i, p_j + \epsilon E_j\} = \delta_{ij} + \epsilon \left(\{F_i, p_j\} + \{q_i, E_j\}\right) + O(\epsilon^2) \\ &= \delta_{ij} + \epsilon \left(\frac{\partial F_i}{\partial q_j} + \frac{\partial E_j}{\partial p_i} \right) + O(\epsilon^2), \\ \{Q_i, Q_j\} &= \{q_i + \epsilon F_i, q_j + \epsilon F_j\} = \epsilon \left(\frac{\partial F_j}{\partial p_i} - \frac{\partial F_i}{\partial p_j} \right) + O(\epsilon^2), \\ \{P_i, P_j\} &= \{p_i + \epsilon E_i, p_j + \epsilon E_j\} = \epsilon \left(\frac{\partial E_i}{\partial q_j} - \frac{\partial E_j}{\partial q_i} \right) + O(\epsilon^2). \end{aligned}$$

So at first order for canonical transformation, we require

$$\begin{cases} \dfrac{\partial F_i}{\partial q_j} + \dfrac{\partial E_j}{\partial p_i} = 0 \\ \dfrac{\partial F_j}{\partial p_i} - \dfrac{\partial F_i}{\partial p_j} = 0 \\ \dfrac{\partial E_i}{\partial q_j} - \dfrac{\partial E_j}{\partial q_i} = 0 \end{cases}$$

This is solved by

$$F_i = \frac{\partial G}{\partial p_i}, \qquad E_i = -\frac{\partial G}{\partial q_i}$$

for some function $G(q,p)$. We say $G(q,p)$ generates the transformation. We have found

$$\begin{cases} Q_i = q_i + \epsilon \{q_i, G\} \\ P_i = p_i + \epsilon \{p_i, G\} \end{cases}$$

In general, suppose we have a one-parameter family of canonical transformations

$$\begin{cases} Q_i = Q_i(q,p,s) \\ P_i = P_i(q,p,s) \end{cases}$$

parametrized by s. Then this would be described by a function $G(q, p, s)$ such that

$$\begin{cases} \dfrac{\partial Q_i}{\partial s} = \{Q_i, G\} \\ \dfrac{\partial P_i}{\partial s} = \{P_i, G\} \end{cases}$$

This has the same form of Hamilton's equations, with the Hamiltonian function replaced by a function G and time replaced by the parameter s. Therefore canonical transformations on the phase space can be viewed as phase flow with respect to an appropriately chosen G.

2.4.3 Time-dependent Canonical Transformation

We next consider time-dependent canonical transformation of the form

$$(q_i, p_i) \longmapsto (Q_i(q, p, t), P_i(q, p, t)),$$

which depends on time t. Being canonical transformation we mean again

$$\{Q_i, P_j\}_{p,q} = \delta_{ij}, \qquad \{Q_i, Q_j\}_{p,q} = \{P_i, P_j\}_{p,q} = 0,$$

and therefore the Poisson bracket is preserved as before:

$$\{-,-\}_{p,q} = \{-,-\}_{P,Q}\,.$$

However, when it has time dependence, it will also change the Hamiltonian function.

Let $\begin{cases} \dot{q}_i = \{q_i, \mathscr{H}\} = \dfrac{\partial \mathscr{H}}{\partial p_i} \\ \dot{p}_i = \{p_i, \mathscr{H}\} = -\dfrac{\partial \mathscr{H}}{\partial q_i} \end{cases}$ be Hamilton's equations in (q, p) coordinates.

Let $\begin{cases} \dot{Q}_i = \{Q_i, \mathscr{H}\} = \dfrac{\partial \mathscr{K}}{\partial P_i} \\ \dot{P}_i = \{P_i, \mathscr{H}\} = -\dfrac{\partial \mathscr{K}}{\partial Q_i} \end{cases}$ be Hamilton's equations in (Q, P) coordinates.

We consider the relationship between $\mathscr{H}$ and $\mathscr{K}$. For simplicity, we discuss the case for $n = 1$, i.e., the phase space is $\mathbb{R}^2$. The phase flow is

$$\begin{cases} (q, p) \text{ coordinates}: & \dot{q} = \{q, \mathscr{H}\}\,,\ \dot{p} = \{p, \mathscr{H}\} \\ (Q, P) \text{ coordinates}: & \dot{Q} = \{Q, \mathscr{K}\}\,,\ \dot{P} = \{P, \mathscr{K}\} \end{cases}$$

Here $(Q = Q(q, p, t), P = P(q, p, t))$ is a canonical transformation, i.e., $\{-,-\}_{p,q} = \{-,-\}_{P,Q}$, so both Poisson brackets are denoted by $\{-,-\}$. Now we compute $\dot{Q}$ in both coordinates.

- In (Q, P, t)-coordinates of the extended phase space,

$$\dot{Q} = \{Q, \mathscr{K}\}\,.$$

- In (q, p, t)-coordinates of the extended phase space,

$$\dot{Q} = \{Q, \mathscr{H}\} + \frac{\partial Q(q,p,t)}{\partial t}.$$

Assume these two evolutions are the same, then

$$\frac{\partial Q(q,p,t)}{\partial t} = \{Q, \mathscr{K} - \mathscr{H}\}.$$

Similarly, we have

$$\frac{\partial P(q,p,t)}{\partial t} = \{P, \mathscr{K} - \mathscr{H}\}.$$

Note that since $(q, p, t) \mapsto (Q(q, p, t), P(q, p, t), t)$ is canonical, we can compute $\{-, -\}$ in either coordinates and the results are the same.

Now we consider the following 2-form[1] on the extended phase space: $dP \wedge dQ$. We can express it in (q, p, t) coordinates. By canonicality, the invariance of Poisson bracket implies

$$dP \wedge dQ = dp \wedge dq + \text{terms involving } dt.$$

The terms involving dt can be computed precisely,

$$\begin{aligned}
dP \wedge dQ &= dp \wedge dq + dP \wedge \left(\frac{\partial Q}{\partial t} dt\right) + \left(\frac{\partial P}{\partial t} dt\right) \wedge dQ \\
&= dp \wedge dq + \left(\frac{\partial Q}{\partial t} dP - \frac{\partial P}{\partial t} dQ\right) \wedge dt \\
&= dp \wedge dq + (\{Q, \mathscr{K} - \mathscr{H}\}\, dP - \{P, \mathscr{K} - \mathscr{H}\}\, dQ) \wedge dt \\
&= dp \wedge dq + \left(\frac{\partial(\mathscr{K} - \mathscr{H})}{\partial P} dP + \frac{\partial(\mathscr{K} - \mathscr{H})}{\partial Q} dQ\right) \wedge dt \\
&= dp \wedge dq + \left(\frac{\partial(\mathscr{K} - \mathscr{H})}{\partial P} dP + \frac{\partial(\mathscr{K} - \mathscr{H})}{\partial Q} dQ + \frac{\partial(\mathscr{K} - \mathscr{H})}{\partial t} dt\right) \wedge dt \\
&= dp \wedge dq + d(\mathscr{K} - \mathscr{H}) \wedge dt.
\end{aligned}$$

Here we have computed $\{-, -\}$ in (Q, P, t) and expressed $\mathscr{H}$ in (Q, P, t) in the fourth line. Thus

$$dP \wedge dQ = dp \wedge dq + d(\mathscr{K} - \mathscr{H}) \wedge dt$$

as 2-forms on the extended phase space. The above process can be reversed, and the similar computation holds for $n > 1$ case. Thus we have proved the following

Proposition 2.4.5. *Let $(q_i, p_i, t) \mapsto (Q_i(q, p, t), P_i(q, p, t), t)$ be a canonical transformation. Let $\mathscr{H}$ and $\mathscr{K}$ be the two functions on extended phase space. Then the following two systems of Hamilton's equations*

$$\begin{cases} \dot{q}_i = \{q_i, \mathscr{H}\} \\ \dot{p}_i = \{p_i, \mathscr{H}\} \end{cases} \quad \text{and} \quad \begin{cases} \dot{Q}_i = \{Q_i, \mathscr{K}\} \\ \dot{P}_i = \{P_i, \mathscr{K}\} \end{cases}$$

[1]If you are not familiar with differential forms, see Section 3.1.2.

define the same phase flow on the extended phase space if and only if

$$\sum_i dp_i \wedge dq_i - d\mathscr{H} \wedge dt = \sum_i dP_i \wedge dQ_i - d\mathscr{K} \wedge dt$$

as 2-forms on the extended phase space.

Remark. See Section 3.4.1 for a geometric interpretation of this proposition.

Observe that

$$\sum_i dp_i \wedge dq_i - d\mathscr{H} \wedge dt = d\left(\sum_i p_i dq_i - \mathscr{H} dt\right),$$

the above equation can be written as

$$d\left(\sum_i p_i dq_i - \mathscr{H} dt - \sum_i P_i dQ_i + \mathscr{K} dt\right) = 0.$$

Locally, this is equivalent to

$$\sum_i p_i dq_i - \mathscr{H} dt - \sum_i P_i dQ_i + \mathscr{K} dt = dF$$

for some function F[2] on the extended phase space. F is called the **generating function** of the canonical transformation of the Hamiltonian system.

Remark. If the canonical transformation does not depend on time, i.e., the transformation is $(q,p) \mapsto (Q(q,p), P(q,p))$. Then we can choose $F = F(Q,P) = F(Q(q,p), P(q,p))$ to be a function on the phase space. In this case

$$\mathscr{K} = \mathscr{H}.$$

We are in the situation as we discussed before.

Let us assume (q_i, Q_i, t) are functionally independent, so we can use them to parameterize the extended phase space and express p and P as $p = p(q_i, Q_i, t)$, $P = P(q_i, Q_i, t)$. Let us write

$$\sum_i p_i dq_i - \mathscr{H} dt - \sum_i P_i dQ_i + \mathscr{K} dt = dF_1.$$

Express F_1 as $F_1(q, Q, t)$, we find

$$p_i = \frac{\partial F_1}{\partial q_i}, \qquad P_i = -\frac{\partial F_1}{\partial Q_i}, \qquad \mathscr{K} - \mathscr{H} = \frac{\partial F_1}{\partial t}.$$

Let us call this $F = F_1$ a type-1 generating function.

[2] This follows from Poincaré Lemma. See Example 3.2.11.

For another case, assume (q_i, P_i, t) are functionally independent. Then we can also express everything in terms of (q_i, P_i, t). Using

$$\sum_i P_i dQ_i = d\left(\sum_i P_i Q_i\right) - \sum_i Q_i dP_i,$$

$$\sum_i p_i dq_i - \mathscr{H} dt + \sum_i Q_i dP_i + \mathscr{K} dt = d\left(F + \sum_i P_i Q_i\right).$$

We write $F = F_2(q, P, t) - \sum_i P_i Q_i$. Then by comparing coefficients, we find

$$p_i = \frac{\partial F_2}{\partial q_i}, \qquad Q_i = \frac{\partial F_2}{\partial P_i}, \qquad \mathscr{K} = \mathscr{H} + \frac{\partial F_2}{\partial t}.$$

Let us call this $F = F_2 - \sum_i P_i Q_i$ a type-2 generating function. We summarize as follows.

Type	Generating function	Relation
1	$F = F_1(q, Q, t)$	$p_i = \frac{\partial F_1}{\partial q_i}$, $P_i = -\frac{\partial F_1}{\partial Q_i}$, $\mathscr{K} = \mathscr{H} + \frac{\partial F_1}{\partial t}$
2	$F = F_2(q, P, t) - \sum_i P_i Q_i$	$p_i = \frac{\partial F_2}{\partial q_i}$, $Q_i = \frac{\partial F_2}{\partial P_i}$, $\mathscr{K} = \mathscr{H} + \frac{\partial F_2}{\partial t}$

We also have two other types of canonical transformations by considering (p, Q, t) and (p, P, t). The idea is similar and we leave the formulation to the reader.

Example 2.4.6. *The following coordinate transformation*

$$Q = tq^2, \qquad P = \frac{p}{2tq}$$

defines a time-dependent canonical transformation.

$$pdq - \mathscr{H} dt - PdQ + \mathscr{K} dt = pdq - \frac{p}{2tq} d(tq^2) + (\mathscr{K} - \mathscr{H})dt = \left(\mathscr{K} - \mathscr{H} - \frac{pq}{2t}\right) dt$$

is exact (i.e. equals to dF for some F) if and only if $\mathscr{K} - \mathscr{H} - \frac{pq}{2t}$ is a function of t only, i.e.,

$$\mathscr{K} = \mathscr{H} + \frac{pq}{2t} + \varphi(t).$$

We can get the same result from the point of view of generating function. Observe that we can express P, Q in terms of (q, P, t). This gives a type-2 generating function $F_2(q, P, t)$.

$$p = 2tqP = \frac{\partial F_2}{\partial q}, \qquad Q = tq^2 = \frac{\partial F_2}{\partial P} \qquad \Longrightarrow \qquad F_2 = tq^2 P + \phi(t).$$

Then $\mathscr{K} = \mathscr{H} + \frac{\partial F_2}{\partial t} = \mathscr{H} + q^2 P + \phi'(t) = \mathscr{H} + \frac{qp}{2t} + \phi'(t)$ (hence $\varphi = \phi'$).

2.5 Hamilton-Jacobi Equation

2.5.1 Extremal Action and Hamilton-Jacobi Equation

We have seen that the equation of motion can be obtained from variation method. Now we change a point of view. Let us fix the initial point (q_0, t_0). Let us assume that for an endpoint $(\hat{q}, \hat{t})$, there is a unique extremal curve γ connecting (q_0, t_0) and $(\hat{q}, \hat{t})$.

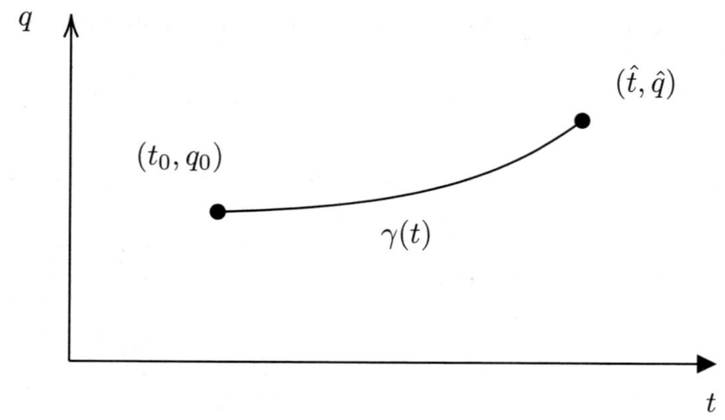

We consider the variation of the endpoint $(\hat{t}, \hat{q})$ in a small neighborhood such that the extremal curve will vary uniquely with respect to the endpoint.

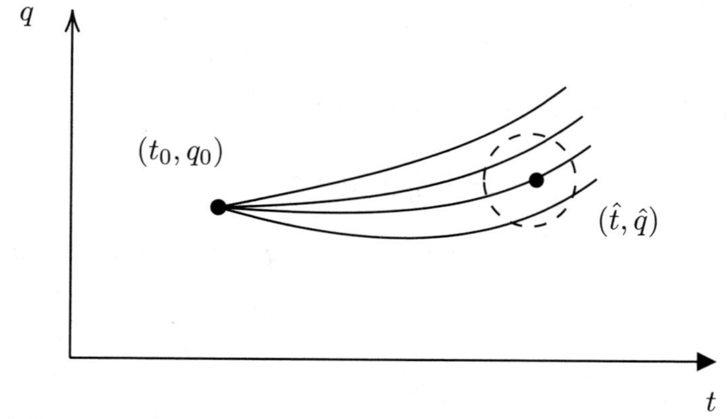

Now we consider the value of the action functional on the extremal curves, viewed as a function on the endpoint denoted by

$$W_{q_0,t_0}(\hat{q}, \hat{t}) = \int_\gamma \mathscr{L} dt,$$

where γ is an extremal curve with $\gamma(t_0) = q_0$, $\gamma(\hat{t}) = \hat{q}$. Since the initial point (q_0, t_0) will be fixed, we will simply write

$$W(\hat{q}, \hat{t}) = \int_\gamma \mathscr{L} dt.$$

We can also lift the curve γ to a curve on the phase space, denoted by $\tilde{\gamma}$.

$$\begin{aligned} \gamma(t) : \quad & q(t) = \gamma,\ \dot{q}(t) = \dot{\gamma}(t) \\ \tilde{\gamma}(t) = (q(t), p(t)) : \quad & q(t) = \gamma(t),\ p(t) = p(q(t), \dot{q}(t)) \end{aligned}$$

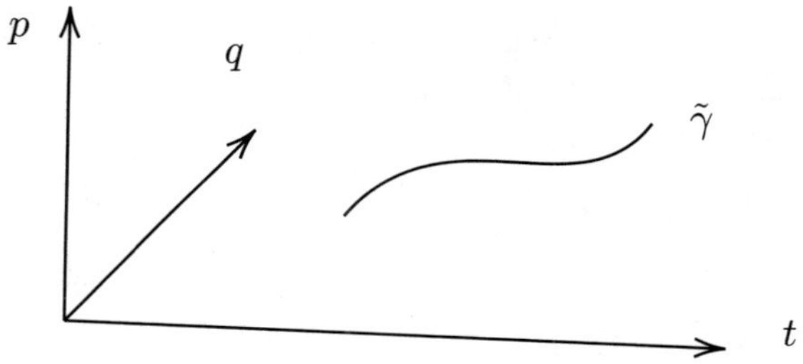

phase space

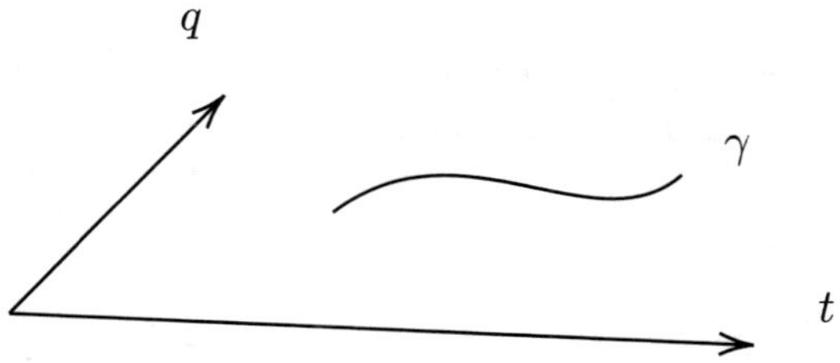

Then we can also write

$$W(\hat{q}, \hat{t}) = \int_{\tilde{\gamma}} pdq - \mathscr{H} dt.$$

Remark. To avoid confusion of notations, we will write $S[\gamma]$ for the action functional on the space of curves, and write $W_{q_0,t_0}(\hat{q}, \hat{t})$ or $W(\hat{q}, \hat{t})$ for the extremal value of the action on extremal curve with endpoint $(\hat{q}, \hat{t})$ (and initial point (q_0, t_0)).

Now we consider the variation of W with respect to $(\hat{q}, \hat{t})$.

① $\frac{\partial W}{\partial \hat{q}}$. Consider the variation of the endpoint

$$(\hat{q}, \hat{t}) \to (\hat{q} + \delta\hat{q}, \hat{t}).$$

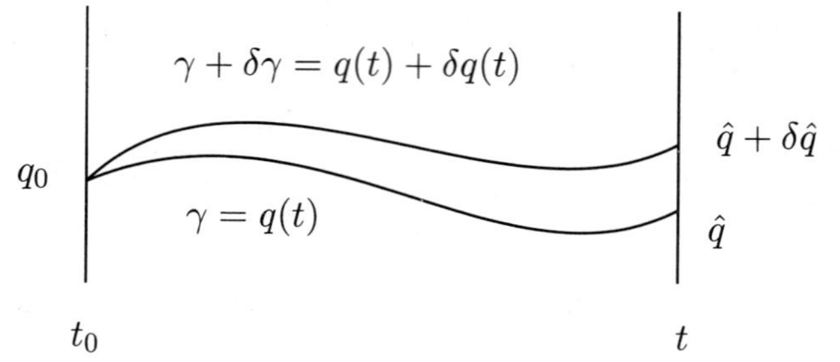

$$\begin{aligned}\delta W = \delta \int_{t_0}^{\hat{t}} \mathscr{L}(q, \dot{q}, t)\, dt &= \sum_i \int_{t_0}^{\hat{t}} \left(\frac{\partial \mathscr{L}}{\partial q_i} \delta q_i + \frac{\partial \mathscr{L}}{\partial \dot{q}_i} (\dot{\delta q_i}) \right) dt \\ &= \sum_i \int_{t_0}^{\hat{t}} \underbrace{\left(\frac{\partial \mathscr{L}}{\partial q_i} - \frac{d}{dt} \frac{\partial \mathscr{L}}{\partial \dot{q}_i} \right)}_{=0 \text{ by EL}} \delta q_i + \sum_i \frac{\partial \mathscr{L}}{\partial \dot{q}_i} \delta q_i \Bigg|_{t_0}^{\hat{t}} \\ &= \sum_i p_i \delta q_i \Bigg|_{t_0}^{\hat{t}} = \sum_i \hat{p}_i \delta \hat{q}_i.\end{aligned}$$

We get

$$\boxed{\frac{\partial W}{\partial \hat{q}_i} = \hat{p}_i}$$

② $\frac{\partial W}{\partial \hat{t}}$. To compute the variation with respect to the endpoint, let us fix an extremal curve γ and consider the variation of W along this γ.

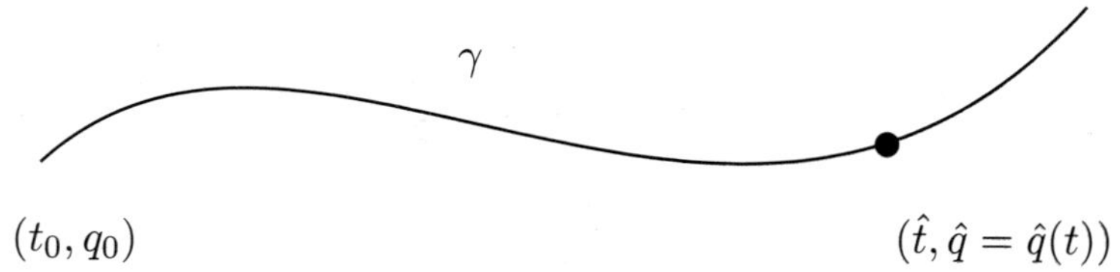

$$W(\hat{q}(\hat{t}), \hat{t}) = \int_{t_0}^{\hat{t}} \mathscr{L}(q, \dot{q}, t)\, dt.$$

Taking the derivative with respect to $\hat{t}$, we find

$$\mathscr{L}(q, \dot{q}, t) = \frac{d}{d\hat{t}} W(\hat{q}(\hat{t}), \hat{t}) = \frac{\partial W}{\partial \hat{q}_i} \dot{\hat{q}}_i + \frac{\partial W}{\partial \hat{t}} = \sum_i \hat{p}_i \dot{\hat{q}}_i + \frac{\partial W}{\partial \hat{t}}.$$

It follows that

$$\frac{\partial W}{\partial \hat{t}} = -\left(\sum_i \hat{p}_i \dot{\hat{q}}_i - \mathscr{L} \right) = -\mathscr{H}.$$

Therefore we find the following variation formula for the extremal value $W(\hat{q},\hat{t})$ of the action

$$\frac{\partial W(\hat{q},\hat{t})}{\partial \hat{q}_i} = \hat{p}_i, \qquad \frac{\partial W(\hat{q},\hat{t})}{\partial \hat{t}_i} = -\mathscr{H}(\hat{q},\hat{p},\hat{t}).$$

To simplify notation, we will simply write (q,t) for the endpoint and write $W = W(q,t)$

$$\frac{\partial W}{\partial q_i} = p_i, \qquad \frac{\partial W}{\partial t} = -\mathscr{H}.$$

Plugging this relation into the expression

$$\mathscr{H} = \mathscr{H}(q,p,t),$$

we have proved

Theorem 2.5.1. *Let $W(q,t)$ be the value of the action on the extremal curve with fixed initial point and endpoint (t,q). Then $W(q,t)$ satisfies the following equation:*

$$\frac{\partial W}{\partial t} + \mathscr{H}\left(q, \frac{\partial W}{\partial q}, t\right) = 0 \qquad (\star)$$

Equation $(\star)$ is called the **Hamilton-Jacobi equation**.

Remark. There is a strong analogy between geometric optics and mechanics, where the surface

$$W = \text{constant}$$

in mechanics plays the role of wavefronts in optics. We will explore this analogy in Section 2.6.

Example 2.5.2. *Consider a particle of mass m moving under the force with potential V.*

$$\mathscr{L} = \frac{1}{2}m\dot{q}^2 - V(q), \qquad \mathscr{H} = \frac{p^2}{2m} + V(q), \qquad p = m\dot{q}.$$

Then the Hamilton-Jacobi equation is

$$\frac{\partial W}{\partial t} + \frac{1}{2m}\left(\frac{\partial W}{\partial q}\right)^2 + V(q) = 0.$$

Remark. It is worthwhile to compare this with the Schrödinger equation

$$i\hbar\frac{\partial \psi}{\partial t} = -\frac{\hbar^2}{2m}\frac{\partial^2 \psi}{\partial x^2} + V(x)\psi$$

in quantum mechanics. Under the change of function

$$\psi = e^{iW/\hbar},$$

the Schrödinger equation is transformed to

$$-\frac{\partial W}{\partial t} = -\frac{i\hbar}{2m}\frac{\partial^2 W}{\partial x^2} + \frac{1}{2m}\left(\frac{\partial W}{\partial x}\right)^2 + V(x).$$

Sending $\hbar \to 0$, this becomes the Hamilton-Jacobi equation above. We see that the $\hbar \to 0$ limit of quantum mechanics becomes classical mechanics. The $\hbar \to 0$ is also called the classical limit.

We illustrate the above ideas by the simple example of free particle moving in $\mathbb{R}^n$.

$$\vec{q} \in \mathbb{R}^n, \qquad V(q) = 0, \qquad \mathscr{H} = \frac{1}{2m}\sum_i p_i^2.$$

The Hamilton-Jacobi equation is

$$\frac{\partial W}{\partial t} + \frac{1}{2m}\sum_i \left(\frac{\partial W}{\partial q_i}\right)^2 = 0.$$

One obvious solution is

$$W = \frac{m(\vec{q} - \vec{q}_0)^2}{2(t - t_0)},$$

where $(t_0, \vec{q}_0)$ is the initial point.

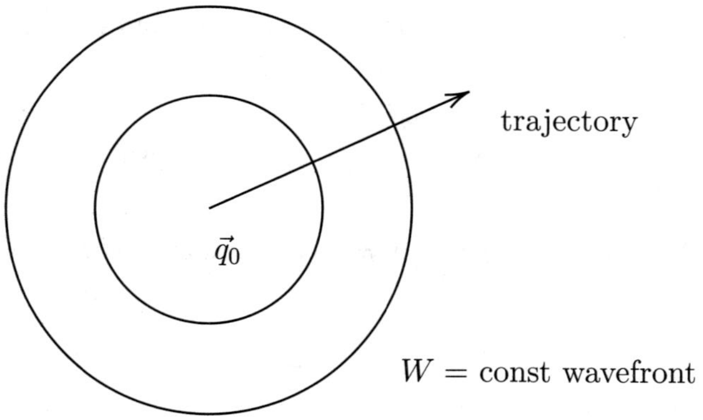

Another solution is

$$W = m\vec{q}\cdot\vec{v} - Et, \qquad \vec{v}, E \text{ are constants.}$$

Plugging into the Hamilton-Jacobi equation, we find

$$E = \frac{1}{2}m\vec{v}^2$$

which is the energy. The solution $W = m\vec{q}\cdot\vec{v} - Et$ is a plane wave. Using $\vec{p} = \frac{\partial W}{\partial \vec{q}} = m\vec{v}$, we see $\vec{v}$ is the velocity as expected, and $\mathscr{H} = -\frac{\partial W}{\partial t} = E$ is the energy. The wavefront looks like

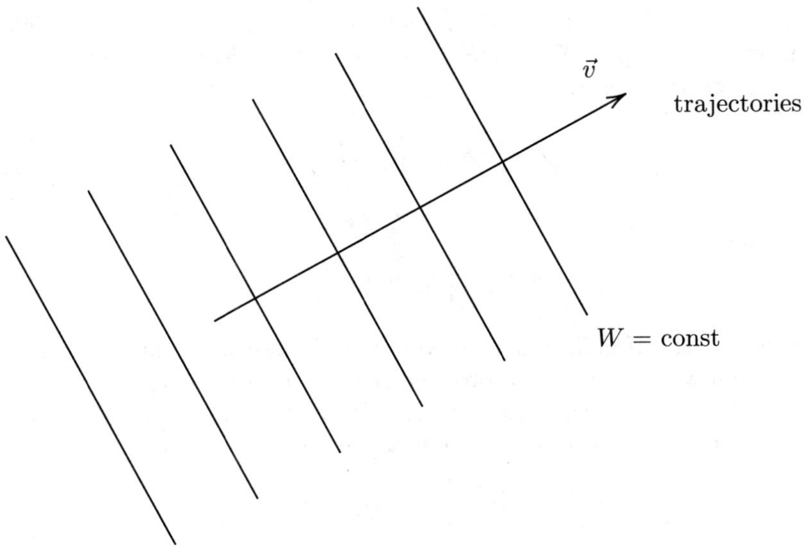

We consider another example of Harmonic oscillator

$$\mathscr{H} = \frac{p^2}{2m} + \frac{kq^2}{2}.$$

The Hamilton-Jacobi equation reads

$$\frac{\partial W}{\partial t} + \frac{1}{2m}\left(\frac{\partial W}{\partial q}\right)^2 + \frac{kq^2}{2} = 0.$$

We look for solutions of the form

$$W = W_0(q) - Et, \qquad \text{where } E = -\frac{\partial W}{\partial t} = \text{Energy}.$$

Then $\frac{dW_0}{dq} = \sqrt{2mE - mkq^2}$, from which we find

$$W_0(q) = \int^q \sqrt{2mE - mkq^2}\, dq = \frac{1}{2} l \sqrt{2mE}\left(\theta + \frac{1}{2}\sin 2\theta\right) + \text{constant}.$$

Here $l = \sqrt{2E/k}$ and $\theta = \arcsin(q/l)$.

2.5.2 Canonical Transformation via Hamilton-Jacobi

Consider the Hamiltonian system

$$(q_i, p_i, \mathscr{H} = \mathscr{H}(q_i, p_i, t)).$$

We try to find a canonical transformation that changes $\mathscr{H}$ to $\mathscr{K}$ which takes a simpler form. The simplest case is $\mathscr{K} = 0$. We consider a type-2 generating function $F_2(q, P, t)$. Then we want to have

$$\mathscr{K} = \mathscr{H}(q, p, t) + \frac{\partial F_2}{\partial t} = 0.$$

Since $p_i = \frac{\partial F_2}{\partial q_i}$, we need

$$\mathscr{H}\left(q, \frac{\partial F_2}{\partial q}, t\right) + \frac{\partial F_2}{\partial t} = 0.$$

This is precisely the Hamilton-Jacobi equation. It indicates that solving Hamilton-Jacobi equation will enable us to solve the Hamiltonian system.

The Hamilton-Jacobi equation has the form of a first-order PDE in $(n+1)$-variables. Suppose there exists a solution of the form

$$F_2 = F_2(q_1, \cdots, q_n; \alpha_1, \cdots, \alpha_{n+1}, t)$$

where the quantities $\alpha_1, \cdots, \alpha_{n+1}$ are $(n+1)$ independent constants. Such solutions are called **complete solutions**. There is an obvious constant by translation and we write

$$F_2 = W(q_1, \cdots, q_n; \alpha_1, \cdots, \alpha_n, t) + \alpha_{n+1}.$$

We use W as a type-2 generating function, and take

$$P_i = \alpha_i$$

being the momentum of the new coordinates. The conjugate coordinates is given by

$$Q_i = \beta_i = \frac{\partial W(q, \alpha, t)}{\partial \alpha_i}.$$

Also $p_i = \frac{\partial W}{\partial q_i}$. Then

$$(q_i, p_i, t) \longmapsto (Q_i = \alpha_i, P_i = \beta_i, t)$$

given a canonical transformation with type-2 generating function W such that the new Hamilton is $\mathscr{K} = 0$. In particular,

$$P_i = \alpha_i, \qquad Q_i = \beta_i \qquad \text{are constants of motion.}$$

Inverting the above relations, we can solve

$$q_i = q_i(\alpha, \beta, t), \qquad p_i = p_i(\alpha, \beta, t).$$

This gives the expression for the corresponding canonical transformation.

Remark. The existence of complete solution of the Hamilton-Jacobi equation is related to the Liouville integrability of the system, which does not hold in general. This relation will be explained in Section 4.3.2.

Example 2.5.3 (Harmonic Oscillator).

$$\mathscr{H} = \frac{1}{2m}(p^2 + m^2\omega^2 q^2), \qquad \omega = \sqrt{k/m}.$$

The Hamilton-Jacobi equation is solved by $W = W_0 - \alpha t$, *where* α *is the energy and*

$$W_0 = \int^q \sqrt{2m\alpha - mkq^2}\, dq.$$

Then the new conjugate coordinate is

$$\beta = \frac{\partial W}{\partial \alpha} = \int \frac{mdq}{\sqrt{2m\alpha - mkq^2}} - t = \frac{1}{\omega} \arcsin\left(q\sqrt{\frac{m\omega^2}{2\alpha}}\right) - t.$$

This allows us to solve

$$q = \sqrt{\frac{2\alpha}{m\omega^2}} \sin\left(\omega t + \omega\beta\right),$$

which is the familiar solution for a harmonic oscillator. The momentum is given by

$$p = \frac{\partial W}{\partial q} = \sqrt{2m\alpha - m^2\omega^2 q^2} = \sqrt{2m\alpha} \cos\left(\omega t + \omega\beta\right).$$

The constants α, β *are related to the initial condition of* q, p *at* $t = 0$.

2.6 Geometric Optics

Many aspects of Hamiltonian mechanics arose from the study of geometric optics. We give a brief discussion on application of methods developed so far to geometric optics. Recall **Fermat's Principle of Least Time**: when a ray of light goes from one point to another, it chooses the path that takes the least time.

2.6.1 Eikonal Equation

The speed of light in a medium is c/n, where c is the speed of light in vacuum, and n is the refraction index. In general, n may vary along the medium, and we will write it as a function

$$n = n(x, y, z),$$

where x, y, z are space coordinates. For a path γ from a point A to a point B in the medium, the time required for the light to travel along γ is

$$\int_\gamma \frac{ds}{c/n} = \frac{1}{c}\int_\gamma nds.$$

Here ds is the arc length element. We take

$$S = \int_\gamma nds$$

as the action functional. The travel path of light is given by the extremal path of S.

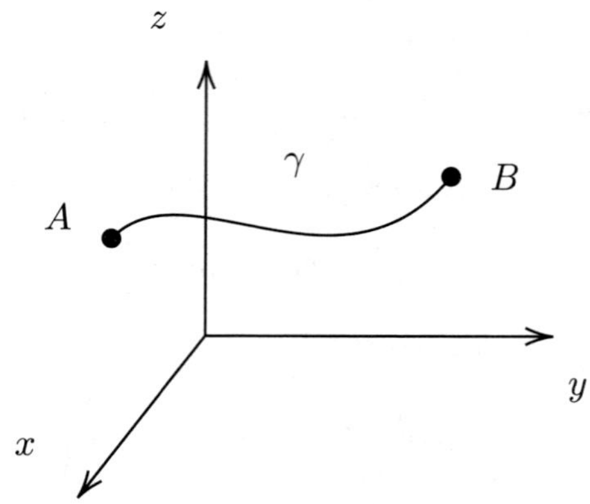

Let us use z as a parameter of the path. Then

$$\gamma(z) = (x(z), y(z), z), \qquad ds = \sqrt{dx^2 + dy^2 + dz^2} = \sqrt{1 + \left(\frac{dx}{dz}\right)^2 + \left(\frac{dy}{dz}\right)^2}\, dz.$$

The action is expressed by

$$S = \int_A^B n(x, y, z)\sqrt{1 + (x')^2 + (y')^2}dz, \qquad \text{where } x' = \frac{dx}{dz},\ y' = \frac{dy}{dz}.$$

In particular, the Lagrangian is

$$\mathscr{L}(x, y, x', y', z) = n(x, y, z)\sqrt{1 + (x')^2 + (y')^2}.$$

The conjugate momentums are given by

$$p_x = \frac{\partial \mathscr{L}}{\partial x'} = \frac{nx'}{\sqrt{1 + (x')^2 + (y')^2}}, \qquad p_y = \frac{\partial \mathscr{L}}{\partial y'} = \frac{ny'}{\sqrt{1 + (x')^2 + (y')^2}}.$$

We can solve x', y' in terms of p_x, p_y. Observe that

$$\sqrt{n^2 - p_x^2 - p_y^2} = \frac{n}{\sqrt{1 + (x')^2 + (y')^2}},$$

therefore

$$x' = \frac{p_x}{\sqrt{n^2 - (p_x)^2 - (p_y)^2}}, \qquad y' = \frac{p_y}{\sqrt{n^2 - (p_x)^2 - (p_y)^2}}.$$

The Hamiltonian is then given by

$$\begin{aligned}\mathscr{H} &= p_x x' + p_y y' - \mathscr{L} \\ &= \frac{p_x^2 + p_y^2}{\sqrt{n^2 - (p_x)^2 - (p_y)^2}} - \frac{n^2}{\sqrt{n^2 - (p_x)^2 - (p_y)^2}} \\ &= -\sqrt{n^2 - (p_x)^2 - (p_y)^2}.\end{aligned}$$

The Hamilton-Jacobi equation is read by

$$\frac{\partial W}{\partial z} - \sqrt{n^2 - \left(\frac{\partial W}{\partial x}\right)^2 - \left(\frac{\partial W}{\partial y}\right)^2} = 0$$

from which we arrive at

$$\left(\frac{\partial W}{\partial x}\right)^2 + \left(\frac{\partial W}{\partial y}\right)^2 + \left(\frac{\partial W}{\partial z}\right)^2 = n^2.$$

This is called the **eikonal equation**, and W is called the **eikonal function**.

Example 2.6.1. *Assume the refraction index n is constant. Then it can be observed that we have a complete solution*

$$W(x, y, \alpha_1, \alpha_2, z) = \alpha_1 x + \alpha_2 y + \sqrt{n^2 - \alpha_1^2 - \alpha_2^2}\, z.$$

We find the canonical transformation

$$P_1 = \alpha_1, \qquad P_2 = \alpha_2,$$

$$\beta_1 = Q_1 = \frac{\partial W}{\partial \alpha_1} = x - \frac{\alpha_1 z}{\sqrt{n^2 - \alpha_1^2 - \alpha_2^2}}, \qquad \beta_2 = Q_2 = \frac{\partial W}{\partial \alpha_2} = y - \frac{\alpha_2 z}{\sqrt{n^2 - \alpha_1^2 - \alpha_2^2}}.$$

Then the solution of light path is solved by

$$(x, y, z) = \left(\beta_1 + \frac{\alpha_1 z}{\sqrt{n^2 - \alpha_1^2 - \alpha_2^2}}, \beta_2 + \frac{\alpha_2 z}{\sqrt{n^2 - \alpha_1^2 - \alpha_2^2}}, z\right).$$

This is a straight line as expected.

2.6.2 Wavefront

Given a time t, we look at the set of all points to which light from a given initial point can travel in time less or equal to t. The boundary of the set is called the **wavefront** of the initial point after time t. It consists of points to which light can travel in time t and not faster. It is clear in the above discussion that the wavefront is described by the Hamilton-Jacobi equation (extremal function of the time $\frac{1}{c}\int nds$).

The solution W of the Hamilton-Jacobi equation gives rise to the wavefront as follows. The velocity is

$$\left(\frac{dx}{dz}, \frac{dy}{dz}, \frac{dz}{dz}\right) = \frac{\left(p_x, p_y, \sqrt{n^2 - (p_x)^2 - (p_y)^2}\right)}{\sqrt{n^2 - (p_x)^2 - (p_y)^2}}.$$

Since

$$p_x = \frac{\partial W}{\partial x}, \qquad p_y = \frac{\partial W}{\partial y}, \qquad \sqrt{n^2 - (p_x)^2 - (p_y)^2} \overset{\text{HJE}}{=\!=} \frac{\partial W}{\partial z},$$

we find

$$\left(\frac{dx}{dz}, \frac{dy}{dz}, \frac{dz}{dz}\right) = \frac{1}{\left(\frac{\partial W}{\partial z}\right)} \left(\frac{\partial W}{\partial x}, \frac{\partial W}{\partial y}, \frac{\partial W}{\partial z}\right),$$

which is proportional to the gradient of W. Thus the light rays intersect orthogonally with the level surface $W(x, y, z)$ =const. Such surfaces constitute the wavefronts.

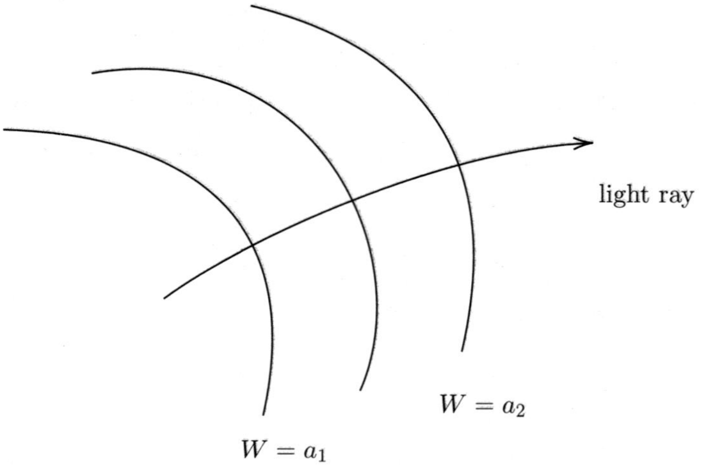

The wavefront $W = a_2$ is obtained from the wavefront $W = a_1$ by the propagation of light ray by a time $(a_2 - a_1)/c$. In fact,

$$\begin{aligned} a_2 - a_1 &= \int_\gamma dW = \int_{z_1}^{z_2} \left(\frac{\partial W}{\partial x}\frac{dx}{dz} + \frac{\partial W}{\partial y}\frac{dy}{dz} + \frac{\partial W}{\partial z}\frac{dz}{dz}\right) dz \\ &= \int_{z_1}^{z_2} \frac{1}{\left(\frac{\partial W}{\partial z}\right)} \left(\left(\frac{\partial W}{\partial x}\right)^2 + \left(\frac{\partial W}{\partial y}\right)^2 + \left(\frac{\partial W}{\partial z}\right)^2\right) dz \\ &= \int_{z_1}^{z_2} \frac{n^2}{\sqrt{n^2 - (p_x)^2 - (p_y)^2}} dz = \int_{z_1}^{z_2} n\sqrt{1 + \left(\frac{dx}{dz}\right)^2 + \left(\frac{dy}{dz}\right)^2} dz \\ &= \int_{z_1}^{z_2} nds. \end{aligned}$$

2.6.3 Maxwell Fisheye

We consider the example where n is a function of $r = \sqrt{x^2 + y^2 + z^2}$ of the form

$$n(r) = \frac{n_0}{1 + (r/a)^2}, \qquad n_0, a \text{ are constants.}$$

This is known as **Maxwell's fisheye**.

It can be shown similarly that in a medium with spherical symmetry, the light ray will lie on a plane passing through the origin. Let us assume that the light ray lies on the xy-plane. Then we are reduced to consider the 2-dim eikonal equation:

$$\left(\frac{\partial W}{\partial x}\right)^2 + \left(\frac{\partial W}{\partial y}\right)^2 = n^2.$$

In terms of polar coordinates

$$(x = r\cos\theta,\ y = r\sin\theta),$$

the eikonal equation becomes

$$\left(\frac{\partial W}{\partial r}\right)^2 + \frac{1}{r^2}\left(\frac{\partial W}{\partial \theta}\right)^2 = n^2.$$

We look for a complete solution of the form

$$W = A(r) + \alpha\theta, \qquad \alpha \text{ is constant.}$$

Plugging this into the above equation, we get

$$A'(r) = \sqrt{n^2 - \frac{\alpha^2}{r^2}}, \qquad A(r) = \int \sqrt{n^2 - \frac{\alpha^2}{r^2}}\,dr.$$

Thus

$$W = \int \sqrt{n^2 - \frac{\alpha^2}{r^2}}\,dr + \alpha\theta.$$

We obtain the canonical transformation with

$$P = \alpha, \qquad Q = \beta = \frac{\partial W}{\partial \alpha} = \theta - \int \frac{\alpha dr}{r\sqrt{n^2r^2 - \alpha^2}}.$$

Substituting $n = n(r) = \frac{n_0}{1+(r/a)^2}$,

$$\theta - \beta = \int \frac{\alpha dr}{r\sqrt{\left(\frac{n_0}{1+(r/a)^2}\right)^2 r^2 - \alpha^2}} \overset{\rho = r/a}{=\!=\!=} \int \frac{\alpha\left(1+\rho^2\right)d\rho}{\rho\sqrt{\rho^2 a^2 n_0^2 - \alpha^2\left(1+\rho^2\right)^2}}$$

$$\overset{K=\frac{\alpha}{an_0}}{=\!=\!=} \int \frac{K\left(1+\rho^2\right)d\rho}{\rho\sqrt{\rho^2 - K^2\left(1+\rho^2\right)^2}} = \arcsin\left(\frac{K}{\sqrt{1-4K^2}}\frac{\rho^2-1}{\rho}\right).$$

We get

$$\boxed{\sin(\theta - \beta) = \frac{\alpha}{\sqrt{a^2n_0^2 - 4\alpha^2}}\frac{r^2 - a^2}{ar}}$$

Here α, β are constants. Put $x = r\cos\theta$, $y = r\sin\theta$, we find

$$y\cos\beta - x\sin\beta = \frac{\alpha}{a\sqrt{a^2n_0^2 - 4\alpha^2}}(x^2 + y^2 - a^2)$$

$$\Longrightarrow \qquad (x + b\sin\beta)^2 + (y - b\cos\beta)^2 = a^2 + b^2$$

where $b = \frac{a}{2\alpha}\sqrt{a^2n_0^2 - 4\alpha^2}$. We see that each ray is a circle.

One remarkable thing about Maxwell fisheye is that "all the rays from an arbitrary point A will meet in a point B on the line joining A to the origin". To see this, let

$$A = (r_0\cos\theta_0, r_0\sin\theta_0).$$

Then $\sin(\theta_0 - \beta) = \frac{\alpha}{\sqrt{a^2n_0^2 - 4\alpha^2}}\frac{r_0^2 - a^2}{ar_0}$. The equation of rays passing through A is described by

$$\frac{r^2 - a^2}{r\sin(\theta - \beta)} = \frac{r_0^2 - a^2}{r_0\sin(\theta_0 - \beta)}.$$

Whatever the β is, there is a solution given by

$$B: r = r_1 = \frac{a^2}{r_0}, \qquad \theta = \theta_1 = \theta_0 + \pi.$$

Note that

$$r = a, \qquad \theta = \beta \text{ or } \theta = \beta + \pi$$

lie on the rays. Thus each ray intersects the circle $r = a$ in opposite points. Here is a picture.

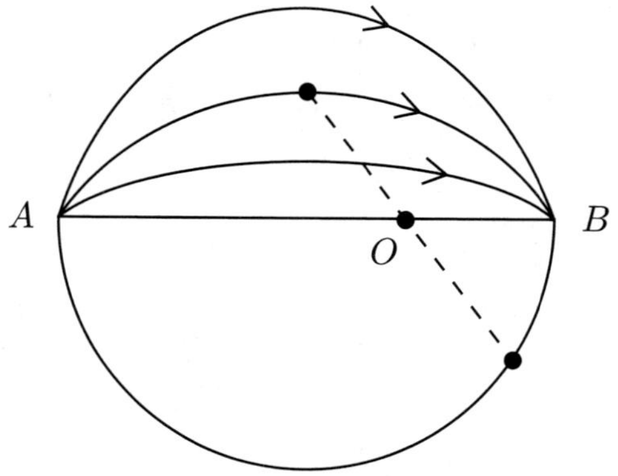

Chapter 3

Interlude of Symplectic Geometry

We give a crash discussion on basic notions in differential geometry and symplectic geometry, with the aim of understanding geometric structures underlying mechanical problems.

3.1 Vector Field and Differential Form

3.1.1 Vector Field

Tangent vector and vector field

Let $\mathcal{U} \in \mathbb{R}^n$ be an open subset and $x \in \mathcal{U}$. We associate an n-dimensional real vector space

$$T_x\mathcal{U} := \{x\} \times \mathbb{R}^n = \{(x, \xi) \mid \xi \in \mathbb{R}^n\}$$

called the **tangent space** of U at x. Elements of $T_x\mathcal{U}$ are called **tangent vectors** to U at x. Geometrically, a tangent vector (x, ξ) indicates a "velocity vector" ξ at the point x.

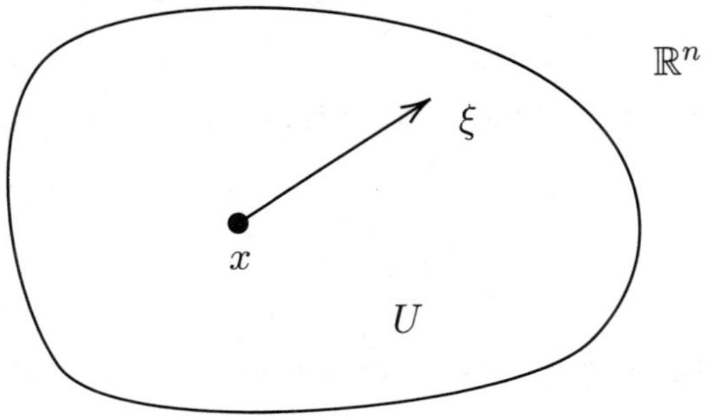

The linear structure is

$$\lambda_1(x,\xi) + \lambda_2(x,\eta) = (x, \lambda_1\xi + \lambda_2\eta), \qquad \lambda_1, \lambda_2 \in \mathbb{R}^n.$$

Example 3.1.1. *Let*

$$\begin{aligned} \gamma : (t_0, t_1) &\longrightarrow \mathcal{U} \subset \mathbb{R}^n \\ t &\longmapsto \gamma(t) \end{aligned}$$

be a curve on $\mathcal{U}$ parameterized by t. Then at each $t \in (t_0, t_1)$, it defines a tangent vector

$$\left(x = \gamma(t),\, \xi = \frac{d\gamma(t)}{dt}\right) \in T_{\gamma(t)}\mathcal{U}.$$

Here $\gamma(t) = (\gamma^1(t), \gamma^2(t), \cdots, \gamma^n(t))$, and

$$\frac{d\gamma}{dt} = \left(\frac{d}{dt}\gamma^1(t), \frac{d}{dt}\gamma^2(t), \cdots, \frac{d}{dt}\gamma^n(t)\right).$$

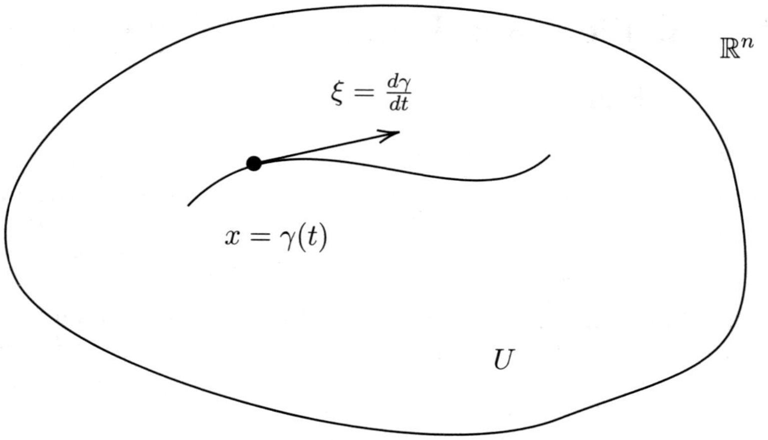

Definition 3.1.2. A **vector field** V on $\mathcal{U}$ is an assignment

$$\begin{aligned} V : \mathcal{U} &\longrightarrow T_x\mathcal{U} \\ x &\longmapsto V(x) = (x, \xi(x)) \end{aligned}$$

V is called continuous/smooth if the components of $\xi(x)$ are continuous/smooth functions on $\mathcal{U}$.

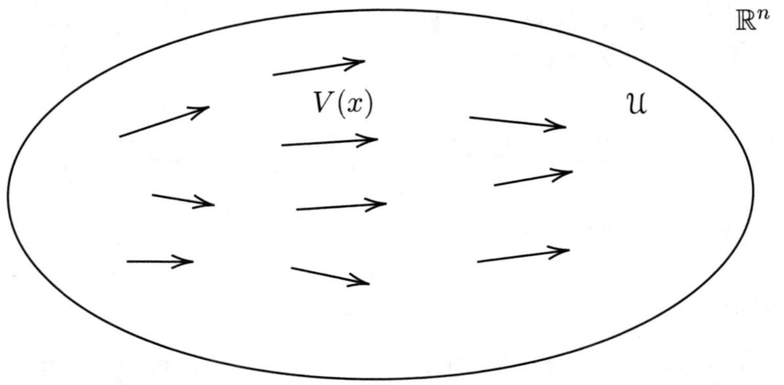

It would be convenient to describe vector fields explicitly in terms of coordinates. Let $x^1, x^2, \cdots, x^n$ be coordinates on $\mathbb{R}^n$. Every point $x \in \mathcal{U}$ is parametrized by

$$x = (x^1, x^2, \cdots, x^n).$$

We denote $\frac{\partial}{\partial x^1}, \cdots, \frac{\partial}{\partial x^n}$ the following vector fields

$$\left.\frac{\partial}{\partial x^1}\right|_x = (x, (1, 0, 0, \cdots, 0)),$$

$$\left.\frac{\partial}{\partial x^2}\right|_x = (x, (0, 1, 0, \cdots, 0)),$$

$$\vdots$$

$$\left.\frac{\partial}{\partial x^n}\right|_x = (x, (0, 0, \cdots, 0, 1)).$$

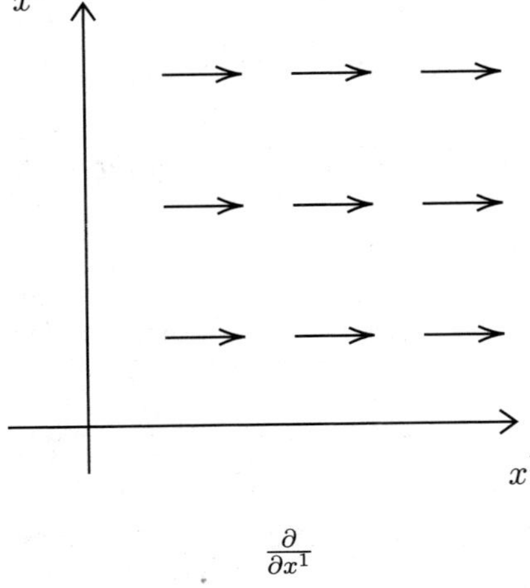

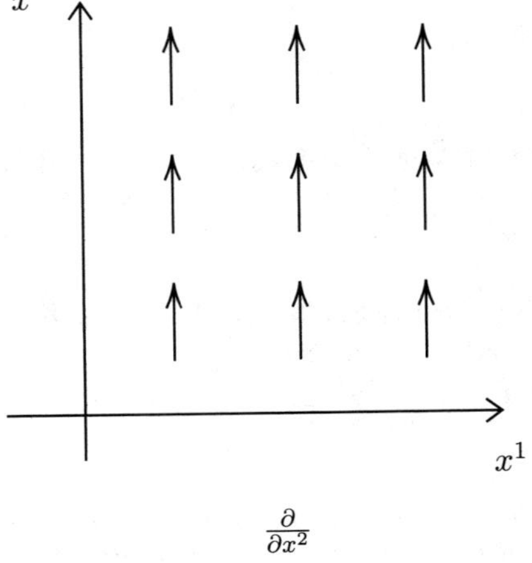

At each $x \in \mathcal{U}$, the tangent vectors $\frac{\partial}{\partial x^1}, \cdots, \frac{\partial}{\partial x^n}$ form a basis of $T_x\mathcal{U}$. Therefore any vector field V can be expressed uniquely as

$$V = V^1(x)\frac{\partial}{\partial x^1} + V^2(x)\frac{\partial}{\partial x^2} + \cdots + V^n(x)\frac{\partial}{\partial x^n} = \sum_{i=1}^{n} V^i(x)\frac{\partial}{\partial x^i},$$

where $V^i(x)$ are functions on $\mathcal{U}$.

Integral curve

We can associate an ordinary differential equation to a vector field V, called the **flow** of the vector field. Intuitively, we can imagine a particle travelling along the direction specified by the vector field at each point. Precisely,

Definition 3.1.3. Let $I \subset \mathbb{R}$ be an open interval. A curve

$$\gamma : I \longrightarrow \mathcal{U}$$

is called an **integral curve** of the vector field V on $\mathcal{U}$ if for each $t \in I$,

$$\frac{d}{dt}\gamma(t) = V(\gamma(t)).$$

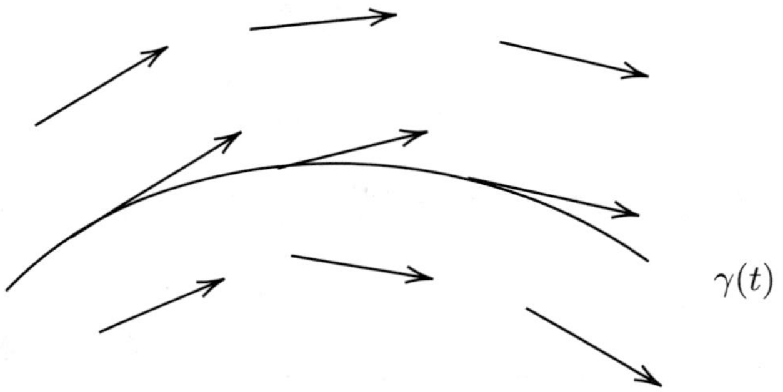

Example 3.1.4. $\gamma(t) = (0, \cdots, 0, \underset{\substack{\uparrow \\ i}}{t}, 0, \cdots, 0)$, $t \in \mathbb{R}$ *is an integral curve of the vector field* $\frac{\partial}{\partial x^i}$.

Theorem 3.1.5. *Let V be a smooth vector field on $\mathcal{U}$. Then for each $x_0 \in \mathcal{U}$, there exists an integral curve*

$$\gamma : (-\varepsilon, \varepsilon) \longrightarrow \mathcal{U}, \qquad \varepsilon \text{ is small}$$

such that $\gamma(0) = x_0$. Moreover, any two such integral curves meeting at x_0 are equal on the intersection of their domain.

This follows from the existence and uniqueness of solutions to the ordinary differential equation

$$\frac{d}{dt}\gamma(t) = V(\gamma(t)).$$

Explicitly, in terms of components $\gamma(t) = (\gamma^1(t), \cdots, \gamma^n(t))$, $V = V^1(x)\frac{\partial}{\partial x^1} + \cdots + V^n(x)\frac{\partial}{\partial x^n}$, the equation is

$$\begin{cases} \frac{d}{dt}\gamma^1(t) = V^1(\gamma^1(t), \cdots, \gamma^n(t)) \\ \frac{d}{dt}\gamma^2(t) = V^2(\gamma^1(t), \cdots, \gamma^n(t)) \\ \vdots \\ \frac{d}{dt}\gamma^n(t) = V^n(\gamma^1(t), \cdots, \gamma^n(t)) \end{cases}$$

with initial condition $\gamma(0) = x_0$.

Example 3.1.6 (Hamiltonian phase flow). *Let $\mathscr{H}(p,q)$ be a Hamiltonian function on the phase space. Then Hamilton's equations describe integral curves for the vector field*

$$\sum_i \left(\frac{\partial \mathscr{H}}{\partial p^i}\frac{\partial}{\partial q^i} - \frac{\partial \mathscr{H}}{\partial q^i}\frac{\partial}{\partial p^i} \right).$$

Differential operator

Given a vector field $V(x) = \sum_i V^i(x)\frac{\partial}{\partial x^i}$, it defines a differential operator acting on smooth functions on $\mathcal{U}$. In fact, given f(x) a smooth function on $\mathcal{U}$, $V(f)$ is the function defined by

$$V(f)(x) = \sum_i V^i(x)\frac{\partial f}{\partial x^i}.$$

It satisfies the following properties:
① $V(c) = 0$ for any constant function c.
② [Leibnitz rule] $V(fg) = V(f)g + fV(g)$.
Actually, any linear map $D : C^\infty(\mathcal{U}) \to C^\infty(\mathcal{U})$ satisfying the above two properties comes from a vector field. This gives an algebraic description of vector fields in terms of the notion of derivatives on the ring of functions.

There is a geometric way to understand the differential operator as follows. Let $f \in C^\infty(\mathcal{U})$. Given $x_0 \in \mathcal{U}$, let $\gamma : (-\varepsilon, \varepsilon) \to \mathcal{U}$ be an integral curve with $\gamma(0) = x_0$. Then

$$V(f)(x_0) = \left.\frac{d}{dt}\right|_{t=0} f(\gamma(t)).$$

In fact,

$$\frac{d}{dt}f(\gamma(t)) = \sum_i \frac{d\gamma^i}{dt}\frac{\partial f}{\partial x^i}(\gamma(t)) = \sum_i V^i(\gamma(t))\frac{\partial f}{\partial x^i}(\gamma(t)).$$

Set $t = 0$, we find

$$\frac{d}{dt}\bigg|_{t=0} f(\gamma(t)) = \sum_i V^i(x_0)\frac{\partial f}{\partial x^i}(x_0) = V(f)(x_0).$$

This formula gives a geometric interpretation of $V(f)$: it describes the infinitesimal change of f along the integral curve of V.

Lie bracket

Let $\mathrm{Vect}(\mathcal{U})$ denote the space of smooth vector fields on $\mathcal{U}$. There is a remarkable Lie algebra structure on $\mathrm{Vect}(\mathcal{U})$ that we now explain.

Let $V = \sum_i V^i(x)\frac{\partial}{\partial x^i}$ and $W = \sum_i W^i(x)\frac{\partial}{\partial x^i}$ be two vector fields on $\mathcal{U}$. Thinking about them as differential operators, we can ask for the new operator

$$[V, W] : C^\infty(\mathcal{U}) \longrightarrow C^\infty(\mathcal{U})$$

given by

$$[V, W]f = V(W(f)) - W(V(f)).$$

Observe that

① $[V, W]c = 0$ if c is a constant function.

② Given any $f, g \in C^\infty(U)$,

$$\begin{aligned}[V, W](fg) &= V(W(fg)) - W(V(fg)) = V(W(f)g + fW(g)) - W(V(f)g + fV(g)) \\ &= V(W(f))g + W(f)V(g) + V(f)W(g) + fV(W(g)) \\ &\quad - W(V(f))g - V(f)W(g) - W(f)V(g) - fW(V(g)) \\ &= ([V, W](f))\, g + f[V, W](g).\end{aligned}$$

As we mentioned above, $[V, W]$ must correspond to a vector field. Explicitly, we find

$$\begin{aligned}[V, W]f &= \sum_{i,j} V^j(x)\frac{\partial}{\partial x^j}\left(W^i(x)\frac{\partial}{\partial x^i}f\right) - \sum_{i,j} W^j(x)\frac{\partial}{\partial x^j}\left(V^i(x)\frac{\partial}{\partial x^i}f\right) \\ &= \sum_i\left(\sum_j V^j(x)\frac{\partial W^i}{\partial x^j} - W^j(x)\frac{\partial V^i}{\partial x^j}\right)\frac{\partial}{\partial x^i}f.\end{aligned}$$

It follows that

$$[V, W] = \sum_i\left(\sum_j V^j(x)\frac{\partial W^i}{\partial x^j} - W^j(x)\frac{\partial V^i}{\partial x^j}\right)\frac{\partial}{\partial x^i} = \sum_i \left(V(W^i) - W(V^i)\right)\frac{\partial}{\partial x^i}.$$

This is called the **Lie bracket** of vector fields.

Proposition 3.1.7. *The Lie bracket* $[-,-]$ *on* $\mathrm{Vect}(\mathcal{U})$ *satisfies*
① $[-,-]$ *is* $\mathbb{R}$*-bilinear.*
② *Skew-symmetry:* $[V,W]=-[W,V], \ \forall V,W\in \mathrm{Vect}(\mathcal{U})$.
③ *Jacobi identity:* $\forall V,W,T\in \mathrm{Vect}(\mathcal{U})$,

$$[V,[W,T]]+[W,[T,V]]+[T,[V,W]]=0.$$

In other words, $(\mathrm{Vect}(\mathcal{U}),[-,-])$ *forms a Lie algebra.*

Proof. Exercise. □

Tangent map

Let $\mathcal{U}_1\in\mathbb{R}^n$ and $\mathcal{U}_2\in\mathbb{R}^m$ be two open subsets. We denote their coordinates by

$$\mathcal{U}_1 : x=(x^1,\cdots,x^n), \qquad \mathcal{U}_2 : y=(y^1,\cdots,y^m).$$

Let $\varphi:\mathcal{U}_1\to\mathcal{U}_2$, $x\mapsto\varphi(x)$ be a smooth map. In coordinates, we write

$$(x^1,\cdots,x^n)\longmapsto(y^1=\varphi^1(x),\cdots,y^n=\varphi^n(x)),$$

where $\varphi^i(x)$ are smooth functions. We can define a corresponding linear map

$$\varphi_* : T_x\mathcal{U}_1\longrightarrow T_{\varphi(x)}\mathcal{U}_2$$

by "**pushing forward**" a tangent vector. In terms of formula, φ_* applied to a basis is

$$\varphi_*\left(\frac{\partial}{\partial x^i}\right)=\sum_{j=1}^m\frac{\partial y^j}{\partial x^i}\frac{\partial}{\partial y^j}=\sum_{j=1}^m\frac{\partial\varphi^j}{\partial x^i}\frac{\partial}{\partial y^j}.$$

This is basically chain rule. In general,

$$\varphi_*\left(\sum_{i=1}^n\xi^i\frac{\partial}{\partial x^i}\right)=\sum_{i=1}^n\sum_{j=1}^m\xi^i\frac{\partial\varphi^j}{\partial x^i}\frac{\partial}{\partial y^j}.$$

Geometrically, let $\gamma:I\to\mathcal{U}_1$ be a curve such that

$$\gamma(0)=x, \qquad \gamma'(0)=\xi=\sum_{i=1}^n\xi^i\frac{\partial}{\partial x^i}.$$

Then $\varphi\circ\gamma:I\to\mathcal{U}_2$ gives a curve on $\mathcal{U}_2$ such that

$$\varphi(\gamma(0))=\varphi(x).$$

Its tangent vector at $t=0$ is

$$\sum_{i=1}^n\left.\frac{d}{dt}\right|_{t=0}\varphi^i(\gamma(t))\frac{\partial}{\partial y^i},$$

whose coefficient can be computed by

$$\left.\frac{d}{dt}\right|_{t=0} \varphi^i(\gamma(t)) = \sum_{j=1}^{n} \left(\left.\frac{d}{dt}\right|_{t=0} \gamma^j(t) \right) \frac{\partial \varphi^i}{\partial x^j}(\gamma(0)) = \sum_{j=1}^{n} \xi^j \frac{\partial \varphi^i}{\partial x^j}(x).$$

We find

$$(\varphi \circ \gamma)'(0) = \varphi_*(\gamma'(0)) = \varphi_*(\xi).$$

Here is the geometric picture of φ_*.

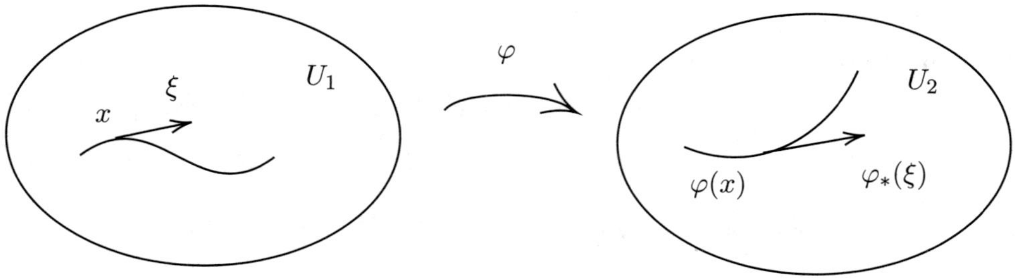

φ_* is called the **tangent map**, or the **push-forward** of tangent vectors.

Proposition 3.1.8. *Let $\mathcal{U}_1 \in \mathbb{R}^n$, $\mathcal{U}_2 \in \mathbb{R}^m$, $\mathcal{U}_3 \in \mathbb{R}^k$ be open subsets. Let $\varphi : \mathcal{U}_1 \to \mathcal{U}_2$, $\phi : \mathcal{U}_2 \to \mathcal{U}_3$ be smooth maps. Then $(\phi \circ \varphi)_* = \phi_* \circ \varphi_*$.*

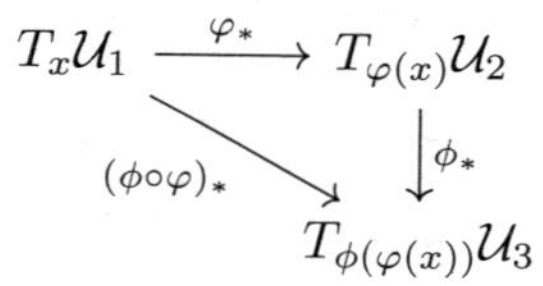

3.1.2 Differential Form

Let $\mathcal{U} \in \mathbb{R}^n$ be an open set. Let $x \in \mathcal{U}$. We have the n-dimensional linear vector space

$$T_x\mathcal{U} = \{\text{tangent vectors at } x\} = \mathrm{Span}_{\mathbb{R}} \left\{ \left.\frac{\partial}{\partial x^1}\right|_x, \cdots, \left.\frac{\partial}{\partial x^n}\right|_x \right\}.$$

Let $T_x^*\mathcal{U}$ denote the linear dual of $T_x\mathcal{U}$:

$$T_x^*\mathcal{U} = \mathrm{Hom}_{\mathbb{R}}(T_x\mathcal{U}, \mathbb{R}).$$

A differential 1-form α on $\mathcal{U}$ is an assignment

$$\begin{aligned} \alpha : \mathcal{U} &\longrightarrow T_x^*\mathcal{U} \\ x &\longmapsto \alpha(x) \end{aligned}$$

α is called continuous/smooth if the components of $\alpha(x)$ are continuous/smooth functions.

Let $\{x^1, \cdots, x^n\}$ be coordinates on $\mathcal{U}$. We write $dx^1, \cdots, dx^n$ be the dual basis of $T_x^*\mathcal{U}$ associated to the basis of $\frac{\partial}{\partial x^1}, \cdots, \frac{\partial}{\partial x^n}$. Let

$$\langle -, - \rangle : T_x^*\mathcal{U} \times T_x\mathcal{U} \longrightarrow \mathbb{R}$$

be the natural pairing via evaluation, then

$$\left\langle dx^i, \frac{\partial}{\partial x^j} \right\rangle = \delta^i_j.$$

In terms of the basis $\{dx^1, \cdots, dx^n\}$, a 1-form α can be expressed as

$$\alpha = \alpha_1(x)dx^1 + \alpha_2(x)dx^2 + \cdots + \alpha_n(x)dx^n.$$

Here $\alpha_1(x)$, $\cdots$, $\alpha_n(x)$ are functions on $\mathcal{U}$. They are continuous/smooth if and only if α is continuous/smooth.

Example 3.1.9. *Let $f(x)$ be a smooth function on $\mathcal{U}$. Then*

$$df := \frac{\partial f}{\partial x^1}dx^1 + \frac{\partial f}{\partial x^2}dx^2 + \cdots + \frac{\partial f}{\partial x^n}dx^n$$

is a smooth 1-form on $\mathcal{U}$.

Let $\Omega^1(\mathcal{U})$ denote the space of smooth 1-forms on $\mathcal{U}$. Then we have a natural pairing

$$\begin{aligned} \Omega^1(\mathcal{U}) \times \mathrm{Vect}(\mathcal{U}) &\longrightarrow C^\infty(\mathcal{U}) \\ (\alpha, v) &\longmapsto \langle \alpha, v \rangle \end{aligned}$$

In coordinates, if $\alpha = \sum_i \alpha_i(x)dx^i$, $V = \sum_i V^i(x)\frac{\partial}{\partial x^i}$, then

$$\langle \alpha, v \rangle = \sum_i \alpha_i(x)V^i(x).$$

Definition 3.1.10. Let $V \in \mathrm{Vect}(\mathcal{U})$. We denote

$$\begin{aligned} \iota_V : \Omega^1(\mathcal{U}) &\longrightarrow C^\infty(\mathcal{U}) \\ \alpha &\longmapsto \langle \alpha, v \rangle \end{aligned}$$

Example 3.1.11. *Let $V \in \mathrm{Vect}(\mathcal{U})$, $f \in C^\infty(\mathcal{U})$. Then*

$$V(f) = \sum_i V^i(x)\frac{\partial f}{\partial x^i} = \iota_V df.$$

Example 3.1.12. *Let $\alpha = \sum_i \alpha_i(x)dx^i$. Then*

$$\alpha_i(x) = \iota_{\frac{\partial}{\partial x^i}} \alpha.$$

Definition 3.1.13. Let $\bigwedge^p T_x^*\mathcal{U}$ denote the p-th exterior product of $T_x^*\mathcal{U}$. A **differential** p**-form** ξ on $\mathcal{U}$ is an assignment

$$\begin{aligned}\xi : \mathcal{U} &\longrightarrow \bigwedge^p T_x^*\mathcal{U} \\ x &\longmapsto \xi(x)\end{aligned}$$

A basis of $\bigwedge^p T_x^*\mathcal{U}$ can be chosen as

$$\left\{dx^{i_1} \wedge dx^{i_2} \wedge \cdots \wedge dx^{i_p}\right\}_{i_1<i_2<\cdots<i_p}.$$

In particular, $\dim_{\mathbb{R}}(\bigwedge^p T_x^*\mathcal{U}) = \binom{n}{p} = \frac{n!}{p!(n-p)!}$. We can express a p-form ξ as

$$\xi = \sum_{i_1<i_2<\cdots<i_p} \xi_{i_1 i_2 \cdots i_p}(x) dx^{i_1} \wedge dx^{i_2} \wedge \cdots \wedge dx^{i_p},$$

where $\xi_{i_1 i_2 \cdots i_p}(x)$ are functions on U. Similarly, ξ is called continuous/smooth if all components $\{\xi_{i_1 i_2 \cdots i_p}(x)\}$ are continuous/smooth.

Using the anti-symmetry property of exterior product

$$dx^i \wedge dx^j = -dx^j \wedge dx^i,$$

we can write

$$\xi = \frac{1}{p!} \sum_{i_1 i_2 \cdots i_p} \xi_{i_1 i_2 \cdots i_p}(x) dx^{i_1} \wedge dx^{i_2} \wedge \cdots \wedge dx^{i_p},$$

where $\xi_{i_1 i_2 \cdots i_p}(x)$ is now defined by all values of indices (not only in increase order) and it is completely anti-symmetric with respect to the permutation of indices.

Definition 3.1.14. We denote

$$\Omega^p(\mathcal{U}) = \{\text{smooth } p\text{-forms on } \mathcal{U}\}, \qquad \Omega^\bullet(\mathcal{U}) = \bigoplus_p \Omega^p(\mathcal{U}).$$

$\alpha \in \Omega^p(\mathcal{U})$ is called even/odd if p is even/odd.

Given a p-form and a q-form, we can define their **wedge product** in terms of the natural exterior product

$$\begin{aligned}\wedge : \Omega^p(\mathcal{U}) \times \Omega^q(\mathcal{U}) &\longrightarrow \Omega^{p+q}(\mathcal{U}) \\ (\alpha, \beta) &\longmapsto \alpha \wedge \beta\end{aligned}$$

Proposition 3.1.15. *The wedge product has the following properties:*
① *Bilinear:* $(\lambda_1\alpha_1+\lambda_2\alpha_2)\wedge\beta = \lambda_1\alpha_1\wedge\beta+\lambda_2\alpha_2\wedge\beta$, *where* $\lambda_i \in \mathbb{R}$, $\alpha_i, \beta_i \in \Omega^\bullet(\mathcal{U})$.
② *Associativity:* $(\alpha \wedge \beta) \wedge \gamma = \alpha \wedge (\beta \wedge \gamma)$.
③ *Graded commutativity:* $\alpha \wedge \beta = (-1)^{pq}\beta \wedge \alpha$, $\alpha \in \Omega^p$, $\beta \in \Omega^q$.

In particular, $\alpha \wedge \alpha = 0$ if α is odd. Note that the relation $dx^i \wedge dx^j = -dx^j \wedge dx^i$ is also about the graded commutativity since dx^i is a 1-form.

Exterior derivative

Recall that the total differential defines a map

$$d : C^\infty(\mathcal{U}) = \Omega^0(\mathcal{U}) \longrightarrow \Omega^1(\mathcal{U})$$
$$f \longmapsto df = \sum_i \frac{\partial f}{\partial x^i} dx^i$$

It satisfies the Leibnitz rule:

$$d(fg) = fdg + gdf.$$

Theorem 3.1.16. *There exists a unique map $d : \Omega^\bullet(U) \to \Omega^\bullet(\mathcal{U})$ satisfying*
① *d is $\mathbb{R}$-linear.*
② $d : \Omega^p(\mathcal{U}) \to \Omega^{p+1}(\mathcal{U})$.
③ *$d : \Omega^0(\mathcal{U}) \to \Omega^1(\mathcal{U})$ is the map $f \mapsto df = \sum_i \frac{\partial f}{\partial x^i} dx^i$.*
④ *Graded Leibnitz rule:*

$$d(\alpha \wedge \beta) = d\alpha \wedge \beta + (-1)^p \alpha \wedge \beta \qquad \text{for } \alpha \in \Omega^p(\mathcal{U}).$$

⑤ $d^2 = 0$.
*d is called the **exterior derivative** on differential forms.*

We will not prove this. Instead, we will follow the above rule to write down the explicit formula for d. Consider

$$\alpha = \frac{1}{p!} \sum_{i_1 i_2 \cdots i_p} \alpha_{i_1 i_2 \cdots i_p}(x) dx^{i_1} \wedge dx^{i_2} \wedge \cdots \wedge dx^{i_p} \in \Omega^p(\mathcal{U}).$$

Using the Leibnitz rule and $d^2 = 0$, we have

$$\begin{aligned} d(dx^{i_1} \wedge dx^{i_2} \wedge \cdots \wedge dx^{i_p}) &= (ddx^1) \wedge (dx^{i_2} \wedge \cdots \wedge dx^{i_p}) - dx^{i_1} \wedge d(dx^{i_2} \wedge \cdots \wedge dx^{i_p}) \\ &= -dx^{i_1} \wedge d(dx^{i_2} \wedge \cdots \wedge dx^{i_p}) = \cdots = 0. \end{aligned}$$

Using Leibnitz rule again and ①, ③, we get

$$\begin{aligned} d\alpha = & \frac{1}{p!} \sum_{i_1 i_2 \cdots i_p} d(\alpha_{i_1 i_2 \cdots i_p}(x)) dx^{i_1} \wedge dx^{i_2} \wedge \cdots \wedge dx^{i_p} \\ & + \frac{1}{p!} \sum_{i_1 i_2 \cdots i_p} \alpha_{i_1 i_2 \cdots i_p}(x) d(dx^{i_1} \wedge dx^{i_2} \wedge \cdots \wedge dx^{i_p}) \end{aligned}$$

$$\Longrightarrow \qquad d\alpha = \frac{1}{p!} \sum_{i_1 i_2 \cdots i_p, k} \frac{\partial \alpha_{i_1 i_2 \cdots i_p}}{\partial x^k} dx^k \wedge dx^{i_1} \wedge \cdots \wedge dx^{i_p}$$

This is the explicit formula of d. It is a good exercise to check directly from this that $d^2 = 0$.

Example 3.1.17. *Let $\alpha = \sum_j \alpha_j(x) dx^j$ be a 1-form. Then*

$$d\alpha = \sum_{i,j} \partial_i \alpha_j dx^i \wedge dx^j = \frac{1}{2} \sum_{i,j} (\partial_i \alpha_j - \partial_j \alpha_i) dx^i \wedge dx^j.$$

Pull-back

Differential form has nice functorial properties with respect to mappings. Let $\mathcal{U}_1 \subset \mathbb{R}^n$ and $\mathcal{U}_2 \subset \mathbb{R}^m$. Let

$$\varphi : \mathcal{U}_1 \longrightarrow \mathcal{U}_2$$

be a smooth map. In coordinates, $\varphi : (x^1, \cdots, x^n) \mapsto (y^1 = \varphi^1(x), \cdots, y^m = \varphi^m(x))$, where $\varphi^i(x)$'s are smooth functions.

Let f be a function on $\mathcal{U}_2$, we can define its **pull-back** $\varphi^*(f)$ to be a function on $\mathcal{U}_1$ via

$$\begin{array}{ccc} \mathcal{U}_1 & \xrightarrow{\ \varphi\ } & \mathcal{U}_2 \\ & \underset{\varphi^*(f)}{\searrow} & \big\downarrow f \\ & & \mathbb{R} \end{array} \qquad \varphi^*(f) = f \circ \varphi$$

In coordinates, $\varphi^*(f)(x) = f(\varphi(x))$. We can naturally extend it to define pull-back of forms

$$\varphi^* : \Omega^p(\mathcal{U}_2) \longrightarrow \Omega^p(\mathcal{U}_1).$$

Given

$$\alpha = \frac{1}{p!} \sum_{i_1 i_2 \cdots i_p} \alpha_{i_1 i_2 \cdots i_p}(y) dy^{i_1} \wedge \cdots \wedge dy^{i_p} \in \Omega^p(\mathcal{U}_2),$$

then

$$\begin{aligned} \varphi^*(\alpha) &= \frac{1}{p!} \sum_{i_1 i_2 \cdots i_p} \varphi^*(\alpha_{i_1 i_2 \cdots i_p}(y)) d(\varphi^*(y^{i_1})) \wedge \cdots \wedge d(\varphi^*(y^{i_p})) \\ &= \frac{1}{p!} \sum_{i_1 i_2 \cdots i_p} \alpha_{i_1 i_2 \cdots i_p}(\varphi(x)) d(\varphi^{i_1}(x)) \wedge \cdots \wedge d(\varphi^{i_p}(x)) \\ &= \frac{1}{p!} \sum_{i_1 \cdots i_p, j_1 \cdots j_p} \alpha_{i_1 i_2 \cdots i_p}(\varphi(x)) \frac{\partial \varphi^{i_1}}{\partial x^{j_1}} \cdots \frac{\partial \varphi^{i_p}}{\partial x^{j_p}} dx^{j_i} \wedge \cdots \wedge dx^{j_p}. \end{aligned}$$

This definition is natural that it preserves all key algebraic structures on differential forms.

Proposition 3.1.18. *The pull-back $\varphi^* : \Omega^\bullet(\mathcal{U}_2) \to \Omega^\bullet(\mathcal{U}_1)$ satisfies*

① *Pull-back on functions: for $f \in \Omega^0(\mathcal{U}_2)$,*

$$\varphi^*(f) = f \circ \varphi.$$

② *Compatible with exterior derivative:*

$$\varphi^*(d\alpha) = d(\varphi^* \alpha).$$

③ *Compatible with wedge product:*

$$\varphi^*(\alpha \wedge \beta) = (\varphi^* \alpha) \wedge (\varphi^* \beta).$$

In fact, it is not hard to see that the above properties uniquely determines the pull-back.

3.1.3 Lie Derivative

Let $V = \sum_i V_i(x)\frac{\partial}{\partial x^i}$ be a vector field on $\mathbb{R}^n$. Assume it generates a flow on $\mathbb{R}^n$ denoted by

$$\begin{aligned}\varphi : (-\varepsilon, \varepsilon) \times \mathbb{R}^n &\longrightarrow \mathbb{R}^n \\ (t, x) &\longmapsto \varphi(t, x).\end{aligned}$$

Here $\varphi(0, x) = x$ and

$$\frac{\partial \varphi(t, x)}{\partial t} = V(\varphi(t, x)).$$

In other words, $\varphi(-, x)$ gives the integral curve of V with initial point x at $t = 0$. Set

$$\begin{aligned}\varphi_t : \mathbb{R}^n &\longrightarrow \mathbb{R}^n \\ x &\longmapsto \varphi(t, x)\end{aligned}$$

φ_t is the diffeomorphism of $\mathbb{R}^n$ obtained by following the flow of V by time t.

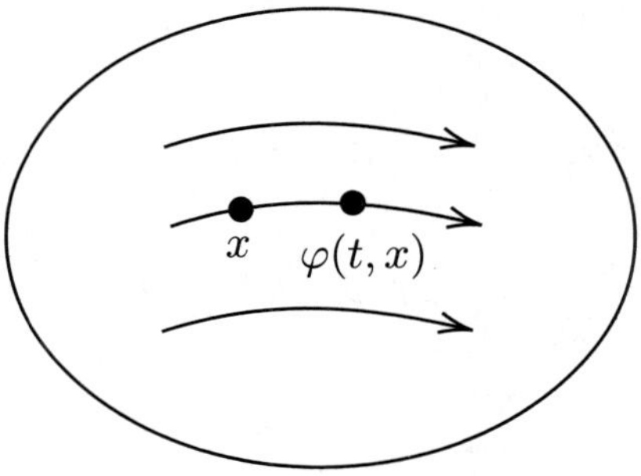

Note that $\varphi_t \circ \varphi_s = \varphi_{t+s}$, $\varphi_t^{-1} = \varphi_{-t}$. This can be visualized from the picture, or proved directly via the uniqueness of the flow equation.

Definition 3.1.19. Let $V \in \mathrm{Vect}(\mathbb{R}^n)$ and $\alpha \in \Omega^p(\mathbb{R}^n)$. We define the Lie derivative

$$\mathscr{L}_V \alpha := \left.\frac{d}{dt}\right|_{t=0} \varphi_t^*(\alpha).$$

This can be viewed as the infinitesimal change of α along the vector field V. To see how it looks like, let $f \in \Omega^0(\mathbb{R}^n)$ be a smooth function.

$$\mathscr{L}_V f = \left.\frac{d}{dt}\right|_{t=0} (\varphi_t^* f) = \left.\frac{d}{dt}\right|_{t=0} f(\varphi_t(x)) = V(f)$$

which is the infinitesimal change along integral curves.

Proposition 3.1.20. *The Lie derivative satisfies the following properties*

① $\mathscr{L}_V$ *commutes with* d*:*

$$\mathscr{L}_V(d\alpha) = d(\mathscr{L}_V \alpha).$$

② *Leibnitz rule:*

$$\mathscr{L}_V(\alpha \wedge \beta) = \mathscr{L}_V(\alpha) \wedge \beta + \alpha \wedge \mathscr{L}_V(\beta).$$

Proof. ① $\mathscr{L}_V(d\alpha) = \frac{d}{dt}\Big|_{t=0} \varphi_t^*(d\alpha) = \frac{d}{dt}\Big|_{t=0} d(\varphi_t^*\alpha) = d\left(\frac{d}{dt}\Big|_{t=0} \varphi_t^*\alpha\right) = d(\mathscr{L}_V\alpha).$

② $\mathscr{L}_V(\alpha \wedge \beta) = \frac{d}{dt}\Big|_{t=0} \varphi_t^*(\alpha \wedge \beta) = \frac{d}{dt}\Big|_{t=0} (\varphi_t^*\alpha) \wedge (\varphi_t^*\beta) = \mathscr{L}_V(\alpha) \wedge \beta + \alpha \wedge \mathscr{L}_V(\beta)$. □

The above properties allows us to write down the formula of $\mathscr{L}_V$ explicitly. In fact, for

$$\alpha = \frac{1}{p!} \sum_{i_1 i_2 \cdots i_p} \alpha_{i_1 i_2 \cdots i_p}(x) dx^{i_1} \wedge dx^{i_2} \wedge \cdots \wedge dx^{i_p},$$

we have

$$\begin{aligned}
\mathscr{L}_V\alpha = & \frac{1}{p!} \sum_{i_1 i_2 \cdots i_p} \mathscr{L}_V(\alpha_{i_1 i_2 \cdots i_p}) dx^{i_1} \wedge dx^{i_2} \wedge \cdots \wedge dx^{i_p} \\
& + \frac{1}{p!} \sum_{i_1 i_2 \cdots i_p} \alpha_{i_1 i_2 \cdots i_p} \mathscr{L}_V(dx^{i_1}) \wedge dx^{i_2} \wedge \cdots \wedge dx^{i_p} \\
& + \cdots + \frac{1}{p!} \sum_{i_1 i_2 \cdots i_p} \alpha_{i_1 i_2 \cdots i_p} dx^{i_1} \wedge dx^{i_2} \wedge \cdots \wedge \mathscr{L}_V(dx^{i_p}) \\
= & \frac{1}{p!} \sum_{i_1 i_2 \cdots i_p} V(\alpha_{i_1 i_2 \cdots i_p}) dx^{i_1} \wedge dx^{i_2} \wedge \cdots \wedge dx^{i_p} \\
& + \frac{1}{p!} \sum_{i_1 i_2 \cdots i_p} \alpha_{i_1 i_2 \cdots i_p} d(\mathscr{L}_V x^{i_1}) \wedge dx^{i_2} \wedge \cdots \wedge dx^{i_p} \\
& + \cdots + \frac{1}{p!} \sum_{i_1 i_2 \cdots i_p} \alpha_{i_1 i_2 \cdots i_p} dx^{i_1} \wedge dx^{i_2} \wedge \cdots \wedge d(\mathscr{L}_V x^{i_p}) \\
= & \frac{1}{p!} \sum_{i_1 i_2 \cdots i_p} V(\alpha_{i_1 i_2 \cdots i_p}) dx^{i_1} \wedge dx^{i_2} \wedge \cdots \wedge dx^{i_p} \\
& + \frac{1}{p!} \sum_{i_1 i_2 \cdots i_p k} \alpha_{i_1 i_2 \cdots i_p} (\partial_k V^{i_1}) dx^k \wedge dx^{i_2} \wedge \cdots \wedge dx^{i_p} \\
& + \cdots + \frac{1}{p!} \sum_{i_1 i_2 \cdots i_p k} \alpha_{i_1 i_2 \cdots i_p} (\partial_k V^{i_p}) dx^{i_1} \wedge dx^{i_2} \wedge \cdots \wedge dx^k.
\end{aligned}$$

Example 3.1.21. *Let* $\Omega = dx^1 \wedge dx^2 \wedge \cdots \wedge dx^n$ *be the standard volume form on* $\mathbb{R}^n$*.*

Let $V = \sum_i V^i(x)\frac{\partial}{\partial x^i}$ *be a vector field. Then*

$$\begin{aligned}\mathscr{L}_V\Omega &= d(V^1(x)) \wedge dx^2 \wedge \cdots \wedge dx^n + dx^1 \wedge d(V^2(x)) \wedge \cdots \wedge dx^n \\ &\quad + \cdots + dx^1 \wedge dx^2 \wedge \cdots \wedge dx^{n-1} \wedge d(V^n(x)) \\ &= \left(\partial_{x^1}V^1 + \partial_{x^2}V^2 + \cdots + \partial_{x^n}V^n\right) dx^1 \wedge dx^2 \wedge \cdots \wedge dx^n \\ &= \operatorname{div}(V)\,\Omega.\end{aligned}$$

Therefore the divergence $\operatorname{div}(V)$ *of the vector field represents infinitesimal change of the volume.*

3.1.4 Stoke's Theorem

Integration of differential forms

Differential forms can be integrated over "oriented" spaces.

Definition 3.1.22. An orientation of an n-dimensional vector space V is the choice of an oriented basis modulo the following equivalence relation: two order basis $e_1, e_2, \cdots, e_n$ and $\tilde{e}_1, \tilde{e}_2, \cdots, \tilde{e}_n$ give the same orientation if

$$e_1 \wedge e_2 \wedge \cdots \wedge e_n = \lambda \tilde{e}_1 \wedge \tilde{e}_2 \wedge \cdots \wedge \tilde{e}_n \in \bigwedge^n V$$

for some $\lambda > 0$.

Here $\bigwedge^n V$ is the n-th exterior product of V, which is a 1-dimensional vector space. In other words, we have

$$\{\text{orientation of } n\text{-dimensional vector space } V\} = \frac{\text{nonzero elements of } \bigwedge^n V}{\text{rescaling by positive factor}}.$$

We see that there are two orientations for V.

Example 3.1.23. *Let* $\{e_1, e_2, e_3\}$ *be the standard basis of* $\mathbb{R}^3$. *Since*

$$e_1 \wedge e_2 \wedge e_3 = -e_2 \wedge e_1 \wedge e_3 = e_2 \wedge e_3 \wedge e_1,$$

we see that the oriented basis

$$e_1, e_2, e_3 \qquad \text{and} \qquad e_2, e_3, e_1$$

give the same orientation, which is different from the orientation by the ordered basis e_2, e_1, e_3.

Let $\mathcal{U} \subset \mathbb{R}^n$ be an open subset. We can define an orientation of $\mathcal{U}$ as a coherent assignment of an orientation for each tangent space $T_x\mathcal{U}$. This assignment has to be continuous on $\mathcal{U}$.

We declare a coordinate system $x^1, x^2, \cdots, x^n$ to be oriented when we specified the orientation of $\mathcal{U}$ by the oriented basis $\frac{\partial}{\partial x^1}, \cdots, \frac{\partial}{\partial x^n}$ at each $T_x\mathcal{U}$. For example, $x^2, x^1, x^3, \cdots, x^n$ gives a different oriented coordinate system than $x^1, x^2, \cdots, x^n$ where the orientation is reversed.

Let $x^1, x^2, \cdots, x^n$ be an oriented coordinate system on $\mathcal{U}$. Let $\alpha \in \Omega^n(\mathcal{U})$ be an n-form. We can write

$$\alpha = f(x)dx^1 \wedge dx^2 \wedge \cdots \wedge dx^n.$$

We define the integral of α over a sufficiently regular subset $K \subset \mathcal{U}$ by

$$\int_K \alpha := \iint \cdots \int_K f(x)dx^1dx^2 \cdots dx^n.$$

Here the right hand side is the multiple integral in standard calculus.

Remark. If we change the orientation of $\mathcal{U}$, say by taking $x^2, x^1, x^3, \cdots, x^n$ as an oriented coordinate system, then the integral of n-form will change a sign. In fact

$$\alpha = f(x)dx^1 \wedge dx^2 \wedge \cdots \wedge dx^n = -f(x)dx^2 \wedge dx^1 \wedge \cdots \wedge dx^n.$$

Then

$$\int_K \alpha = \iint \cdots \int_K (-f(x))dx^2dx^1 \cdots dx^n = -\iint \cdots \int_K f(x)dx^1dx^2 \cdots dx^n.$$

The integral of an n-form α is independent of choice of coordinates for the fixed oriented coordinate system. Let $y^1, y^2, \cdots, y^n$ be another oriented coordinate system on $\mathcal{U}$. Then

$$dx^1 \wedge dx^2 \wedge \cdots \wedge dx^n = \det\left(\frac{\partial x^i}{\partial y^j}\right) dy^1 \wedge dy^2 \wedge \cdots \wedge dy^n$$

with $\det\left(\frac{\partial x^i}{\partial y^j}\right) > 0$ since both coordinates define the same orientation. We can write α in new coordinates

$$\alpha = f(x)dx^1 \wedge dx^2 \wedge \cdots \wedge dx^n = f(x(y)) \det\left(\frac{\partial x^i}{\partial y^j}\right) dy^1 \wedge \cdots \wedge dy^n.$$

In coordinates $\{x^i\}$, we have

$$\int_K \alpha = \iint \cdots \int_K f(x)dx^1dx^2 \cdots dx^n.$$

In coordinates $\{y^i\}$, we have

$$\int_K \alpha = \iint \cdots \int_K f(x(y)) \det\left(\frac{\partial x^i}{\partial y^j}\right) dy^1dy^2 \cdots dy^n.$$

These two expressions are equal by the standard rule of change of variables in multiple integrals.

The above consideration can be generalized to the behavior of integrals under diffeomorphisms. Let $\mathcal{U} \subset \mathbb{R}^n$ with oriented coordinate system $x^1, x^2, \cdots, x^n$ and $\mathcal{V} \subset \mathbb{R}^n$ with oriented coordinate system $y^1, y^2, \cdots, y^n$. Let $\varphi : \mathcal{U} \to \mathcal{V}$ be a diffeomorphism, i.e., φ is smooth, bijective and φ^{-1} is also smooth. Assume φ is orientation-preserving. This means

$$\varphi^*(dy^1 \wedge dy^2 \wedge \cdots \wedge dy^n) = \det\left(\frac{\partial \varphi^i}{\partial x^j}\right) dx^1 \wedge dx^2 \wedge \cdots \wedge dx^n,$$

where $\det\left(\frac{\partial \varphi^i}{\partial x^j}\right) > 0$. Let $K \subset U$ and $\alpha \in \Omega^n(\mathcal{V})$. Then

$$\int_K \varphi^* \alpha = \int_{\varphi(K)} \alpha.$$

This follows again from the rule of change of variables in multiple integrals.

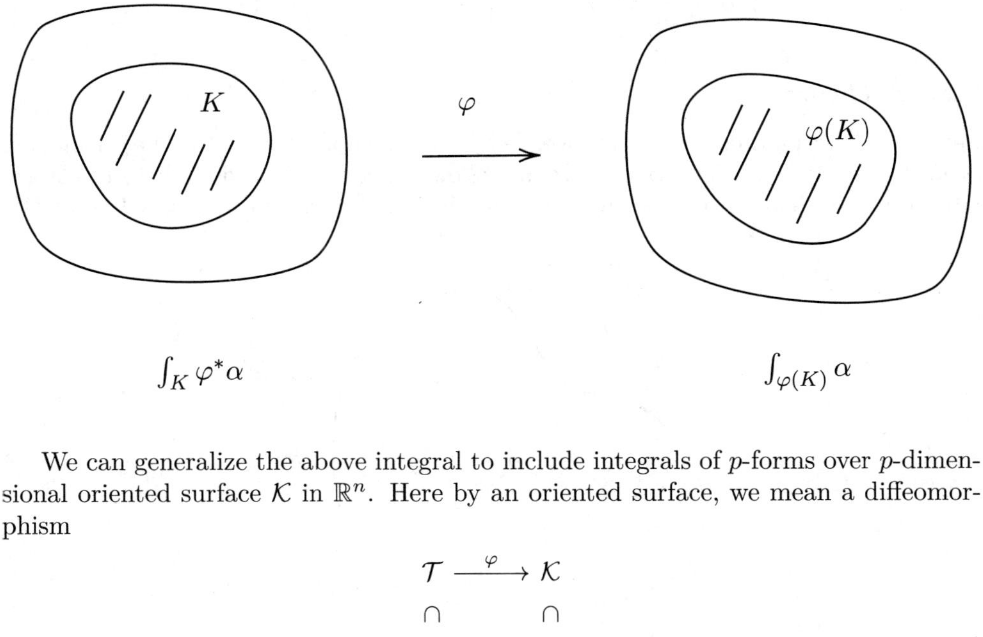

We can generalize the above integral to include integrals of p-forms over p-dimensional oriented surface $\mathcal{K}$ in $\mathbb{R}^n$. Here by an oriented surface, we mean a diffeomorphism

$$\begin{array}{ccc} \mathcal{T} & \xrightarrow{\ \varphi\ } & \mathcal{K} \\ \cap & & \cap \\ \mathbb{R}^p & & \mathbb{R}^n \end{array}$$

Here $\mathcal{T}$ is an oriented region in $\mathbb{R}^p$. Then we define

$$\int_{\varphi(\mathcal{K})} \alpha = \int_{\mathcal{T}} \varphi^* \alpha$$

for $\alpha \in \Omega^p(\mathbb{R}^n)$.

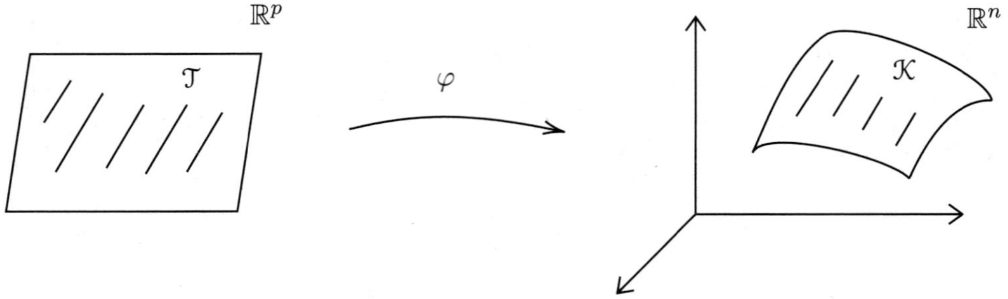

Example 3.1.24. *Let* $\alpha = \sum_i \alpha_i(x)dx^i$ *be a* 1*-form. Let*

$$\gamma : [a,b] \longrightarrow \mathbb{R}^n$$
$$t \longmapsto \gamma(t) = (\gamma^1(t), \gamma^2(t), \cdots, \gamma^n(t))$$

be a smooth curve inside $\mathbb{R}^n$*. Then*

$$\int_\gamma \alpha = \int_a^b \gamma^*(\alpha) = \int_a^b \left(\sum_i \alpha_i(\gamma(t)) \frac{d\gamma^i(t)}{dt}\right) dt.$$

Stokes' Theorem

Let $\mathcal{K} \subset \mathbb{R}^n$ be a p-dimensional surface. Let $\partial\mathcal{K}$ be the boundary of $\mathcal{K}$. The orientation of $\mathcal{K}$ determines an orientation of $\partial\mathcal{K}$ as follows. Let $x \in \mathcal{K}$ and $\vec{n}_x$ be an outward normal vector field on $\partial\mathcal{K}$, i.e., $\vec{n}_x$ is tangent at $\mathcal{K}$ and directed to the exterior of $\mathcal{K}$.

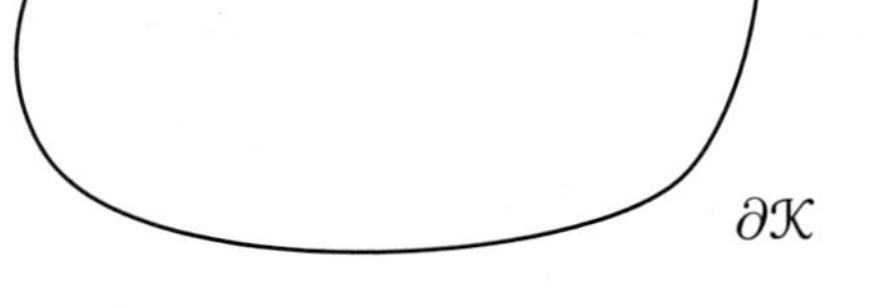

Then an oriented parametrization $\tau^1, \cdots, \tau^{p-1}$ of $\partial\mathcal{K}$ is such that $\vec{n}_x, \frac{\partial}{\partial\tau^1}, \cdots, \frac{\partial}{\partial\tau^{p-1}}$ gives an orientation of $T_x\mathcal{K}$. In terms of this induced orientation, we have the following

Theorem 3.1.25 (Stokes' Theorem)**.**

$$\int_{\mathcal{K}} d\alpha = \int_{\partial\mathcal{K}} \alpha.$$

Here α *is a smooth* $(p-1)$*-form.*

Example 3.1.26. *Consider the case $n = p = 2$. Let*

$$\alpha = \alpha_1(x)dx^1 + \alpha_2(x)dx^2. \qquad d\alpha = \left(\frac{\partial \alpha_2}{\partial x^1} - \frac{\partial \alpha_1}{\partial x^2}\right) dx^1 \wedge dx^2.$$

Then

$$\iint_{\mathcal{K}} \left(\frac{\partial \alpha_2}{\partial x^1} - \frac{\partial \alpha_1}{\partial x^2}\right) dx^1 \wedge dx^2 = \int_{\partial \mathcal{K}} \alpha_1(x)dx^1 + \alpha_2(x)dx^2.$$

This gives Green's Theorem.

Example 3.1.27. *Consider the case $n = p = 3$. Let*

$$\alpha = \alpha_1(x)dx^2 \wedge dx^3 + \alpha_2(x)dx^3 \wedge dx^1 + \alpha_3(x)dx^1 \wedge dx^2.$$

$$d\alpha = \left(\frac{\partial \alpha_1}{\partial x^1} + \frac{\partial \alpha_2}{\partial x^2} + \frac{\partial \alpha_3}{\partial x^3}\right) dx^1 \wedge dx^2 \wedge dx^3.$$

Then

$$\iiint_K \left(\frac{\partial \alpha_1}{\partial x^1} + \frac{\partial \alpha_2}{\partial x^2} + \frac{\partial \alpha_3}{\partial x^3}\right) dx^1 \wedge dx^2 \wedge dx^3$$
$$= \iint_{\partial K} \alpha_1(x)dx^2 \wedge dx^3 + \alpha_2(x)dx^3 \wedge dx^1 + \alpha_3(x)dx^1 \wedge dx^2.$$

This is the integral theorem of Gauss.

As an application of Stokes' Theorem, consider two regions $\mathcal{K}_1$ and $\mathcal{K}_2$ which are the boundary of a region $\mathcal{M}$ as illustrated in the figure below.

$$\partial \mathcal{M} = \mathcal{K}_2 - \mathcal{K}_1.$$

Here the $\pm$ sign is about the orientation.

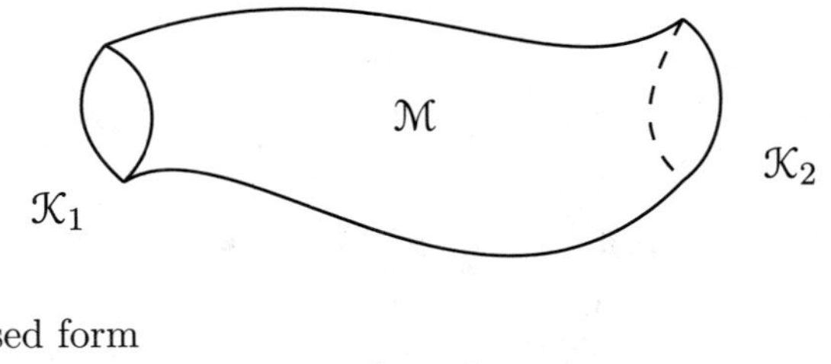

Let α be a closed form

$$d\alpha = 0.$$

Then

$$\int_{\mathcal{K}_2} \alpha - \int_{\mathcal{K}_1} \alpha = \int_{\partial \mathcal{M}} \alpha = \int_{\mathcal{M}} d\alpha = 0.$$

So

$$\int_{\mathcal{K}_1} \alpha = \int_{\mathcal{K}_2} \alpha.$$

In particular, $\int_{\mathcal{K}} \alpha$ will be invariant under any smooth deformation of $\mathcal{K}$ if $d\alpha = 0$.

Example 3.1.28. *Let $\alpha \in \Omega^1(\mathbb{R}^n)$ and $d\alpha = 0$. Let $p, q \in \mathbb{R}^n$ be two points. Then we have*

$$\int_p^q \alpha = \int_\gamma \alpha$$

by choosing any smooth path γ connecting p and q. It does not depend on the choice of γ since any such path γ_1 can be smoothly deformed to another such path γ_2. Then

$$\int_{\gamma_1} \alpha = \int_{\gamma_2} \alpha \qquad \text{by Stokes' Theorem.}$$

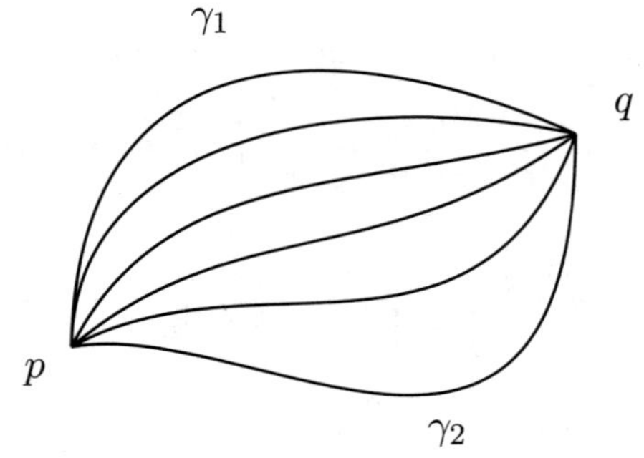

Example 3.1.29. *Consider $\mathcal{U} = \mathbb{R}^2 - \{0\}$ and*

$$\alpha = \frac{xdy - ydx}{x^2 + y^2}, \qquad d\alpha = 0.$$

We consider two paths from $p = -1$ to $q = 1$ by following the upper and lower semicircle.

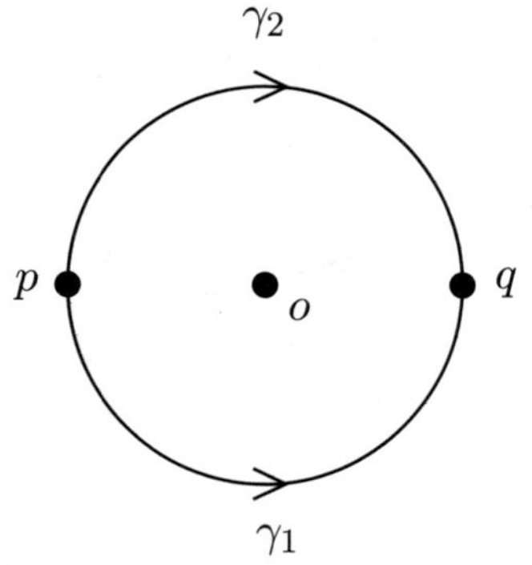

Since there is a hole in the middle, γ_1 can not be deformed to γ_2 inside the region $\mathcal{U}$. This can be also seen by computing

$$\int_{\gamma_1} \alpha - \int_{\gamma_2} \alpha = \oint_C \alpha.$$

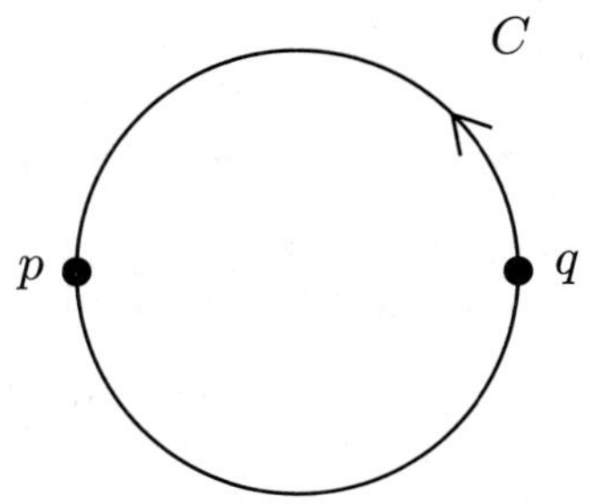

In terms of polar coordinates $x = r\cos\theta$, $y = r\sin\theta$,

$$\alpha = d\theta, \qquad \oint_C \alpha = \int_0^{2\pi} d\theta = 2\pi \neq 0.$$

3.2 Cartan Formula and Poincaré Lemma

3.2.1 Cartan Formula

Recall the derivative $\mathscr{L}_V$ is completely determined by

① $\mathscr{L}_V(f) = V(f)$, $\forall f \in \Omega^0(\mathcal{U}) = C^\infty(\mathcal{U})$.

② $\mathscr{L}_V d = d\mathscr{L}_V$.

③ $\mathscr{L}_V(\alpha \wedge \beta) = (\mathscr{L}_V \alpha) \wedge \beta + \alpha \wedge \mathscr{L}_V(\beta)$, $\forall \alpha, \beta \in \Omega^\bullet(U)$.

In fact, the above three properties uniquely determine the operator $\mathscr{L}_V$: we have

$$\begin{aligned}\mathscr{L}_V(\sum \alpha_{i_1\cdots i_p} dx^{i_1} \wedge \cdots \wedge dx^{i_p}) &= \sum V(\alpha_{i_1\cdots i_p}) dx^{i_1} \wedge \cdots \wedge dx^{i_p} \\ &+ \sum \alpha_{i_1\cdots i_p} d(V(x^{i_1})) \wedge \cdots \wedge dx^{i_p} \\ &+ \cdots + \sum \alpha_{i_1\cdots i_p} dx^{i_1} \wedge \cdots \wedge d(V(x^{i_p})).\end{aligned}$$

Definition 3.2.1. We define the **interior product**

$$\begin{aligned}\iota : \mathrm{Vect}(\mathcal{U}) \times \Omega^p(\mathcal{U}) &\longrightarrow \Omega^{p-1}(\mathcal{U}) \\ (V, \alpha) &\longmapsto \iota_V \alpha\end{aligned}$$

by

$$\begin{aligned}\iota_V(\sum \alpha_{i_1\cdots i_p} dx^{i_1} \wedge \cdots \wedge dx^{i_p}) &= \sum \alpha_{i_1\cdots i_p} (\iota_V dx^{i_1}) \wedge \cdots \wedge dx^{i_p} \\ &- \sum \alpha_{i_1\cdots i_p} dx^{i_1} \wedge (\iota_V dx^{i_2}) \wedge \cdots \wedge dx^{i_p} \\ &+ \cdots + (-1)^{p-1} \sum \alpha_{i_1\cdots i_p} dx^{i_1} \wedge \cdots \wedge (\iota_V dx^{i_p}),\end{aligned}$$

where for $V = \sum V^i(x) \frac{\partial}{\partial x^i}$, we have $\iota_V dx^i = V^i(x)$.

Proposition 3.2.2. *The interior product satisfies the following properties:*

① ι_V *is* $C^\infty(\mathcal{U})$*-linear:* $\iota_V(f\alpha) = f\iota_V\alpha$.

② $\iota_V(df) = V(f)$.
③ $\iota_V(\alpha \wedge \beta) = (\iota_V \alpha) \wedge \beta + (-1)^p \alpha \wedge (\iota_V \beta)$ *for* $\alpha \in \Omega^p(\mathcal{U})$.
Moreover, the above three properties uniquely determines the map ι_V.

Proof. We prove ②: $\iota_V(df) = \iota_V(\sum_i \frac{\partial f}{\partial x^i} dx^i) = \sum_i V^i(x) \frac{\partial f}{\partial x^i} = V(f)$. The uniqueness follows easily by writing down the explicit formula. □

So far we have three operators defined on differential forms

$$\begin{aligned} d : \Omega^p &\longrightarrow \Omega^{p+1} && \text{exterior derivative} \\ \mathscr{L}_V : \Omega^p &\longrightarrow \Omega^{p+1} && \text{Lie derivative} \\ \iota_V : \Omega^p &\longrightarrow \Omega^{p-1} && \text{interior product} \end{aligned}$$

There is a remarkable relationship between these operators, called **Cartan homotopy formula** or **Cartan magic formula**.

Proposition 3.2.3 (Cartan formula). *For any* $V \in \mathrm{Vect}(\mathcal{U})$,

$$\mathscr{L}_V = d\iota_V + \iota_V d \qquad on\ \Omega^\bullet(\mathcal{U}).$$

Proof. Let $D_V = d\iota_V + \iota_V d$. We need to show $\mathscr{L}_V = D_V$. We show that D_V satisfies the following three properties:
① $D_V(f) = V(f)$ for $f \in \Omega^0(U)$.
② $D_V d = dD_V$.
③ $D_V(\alpha \wedge \beta) = (D_V\alpha) \wedge \beta + \alpha \wedge D_V \beta$, $\forall \alpha, \beta \in \Omega^\bullet(U)$.
Since these properties uniquely determine $\mathscr{L}_V$, it will follow that $D_V = \mathscr{L}_V$. Let us now check:
① $D_V(f) = (d\iota_V + \iota_V d)(f) = \iota_V df = V(f)$.
② $D_V d = (d\iota_V + \iota_V d)d = d\iota_V d = d(d\iota_V + \iota_V d) = dD_V$ since $d^2 = 0$.
③ Assume $\alpha \in \Omega^p(U)$, $\beta \in \Omega^q(\mathcal{U})$.

$$\begin{aligned} D_V(\alpha \wedge \beta) &= d\iota_V(\alpha \wedge \beta) + \iota_V d(\alpha \wedge \beta) \\ &= d((\iota_V\alpha) \wedge \beta + (-1)^p \alpha \wedge \iota_V \beta) + \iota_V(d\alpha \wedge \beta + (-1)^p \alpha \wedge d\beta) \\ &= (d\iota_V \alpha) \wedge \beta + (-1)^{p-1}(\iota_V \alpha) \wedge d\beta + (-1)^p d\alpha \wedge \iota_V \beta + \alpha \wedge d\iota_V \beta \\ &\qquad + (\iota_V d\alpha) \wedge \beta + (-1)^{p+1} d\alpha \wedge \iota_V \beta + (-1)^p \iota_V \alpha \wedge d\beta + \alpha \wedge \iota_V d\beta \\ &= (d\iota_V \alpha + \iota_V d\alpha) \wedge \beta + \alpha \wedge (d\iota_V \alpha + \iota_V d)\beta \\ &= D_V \alpha \wedge \beta + \alpha \wedge D_V \beta. \end{aligned}$$

□

The above relation can be further extended to

Theorem 3.2.4. *Let* $V, W \in \mathrm{Vect}(\mathcal{U})$. *Then as operators on* $\Omega^\bullet(\mathcal{U})$,
① $[\mathscr{L}_V, \mathscr{L}_W] := \mathscr{L}_V \mathscr{L}_W - \mathscr{L}_W \mathscr{L}_V = \mathscr{L}_{[V,W]}$.
② $\iota_V \iota_W + \iota_W \iota_V = 0$.
③ $[\mathscr{L}_V, d] := \mathscr{L}_V d - d\mathscr{L}_V = 0$.
④ $\iota_V d + d\iota_V = \mathscr{L}_V$.
⑤ $[\mathscr{L}_V, \iota_W] := \mathscr{L}_V \iota_W - \iota_W \mathscr{L}_V = \iota_{[V,W]}$.

Proof. The idea of proof is very similar to the above proof of Cartan formula. We prove ①, ② for illustration.

① Let $D = [\mathscr{L}_V, \mathscr{L}_W]$. We show that it behaves the same as $\mathscr{L}_{[V,W]}$. Indeed

- $D(f) = (\mathscr{L}_V \mathscr{L}_W - \mathscr{L}_W \mathscr{L}_V)f = V(W(f)) - W(V(f)) = [V, W](f)$.
- $Dd = dD$ since $\mathscr{L}_V$, $\mathscr{L}_W$ both commute with d.
- $D(\alpha \wedge \beta) = D\alpha \wedge \beta + \alpha \wedge D\beta$ can be checked directly.

② This is a simple but important property of interior product. We can check on a basis:

$$\begin{aligned}
&\iota_V \iota_W (dx^{i_1} \wedge \cdots \wedge dx^{i_p}) \\
&= \iota_V \left(\sum_{k=1}^{p} (-1)^{k-1} W^{i_k}(x) dx^{i_1} \wedge \cdots \wedge \widehat{dx^{i_k}} \wedge \cdots \wedge dx^{i_p} \right) \\
&= \sum_{j<k} (-1)^{j+k} V^{i_j}(x) W^{i_k}(x) dx^{i_1} \wedge \cdots \wedge \widehat{dx^{i_j}} \wedge \cdots \wedge \widehat{dx^{i_k}} \wedge \cdots \wedge dx^{i_p} \\
&\qquad - \sum_{k<j} (-1)^{j+k} V^{i_j}(x) W^{i_k}(x) dx^{i_1} \wedge \cdots \wedge \widehat{dx^{i_k}} \wedge \cdots \wedge \widehat{dx^{i_j}} \wedge \cdots \wedge dx^{i_p} \\
&= -\iota_W \iota_V (dx^{i_1} \wedge \cdots \wedge dx^{i_p}).
\end{aligned}$$

□

3.2.2 Poincaré Lemma

Definition 3.2.5. A differential form α is called **closed** if

$$d\alpha = 0.$$

α is called **exact** if there exists β such that

$$\alpha = d\beta.$$

Since $d^2 = 0$, every exact form is closed.

Definition 3.2.6. Let $\mathcal{U} \in \mathbb{R}^n$ be an open subset. We define the p-th de Rham cohomology by

$$H^p_{\mathrm{dR}}(\mathcal{U}) := \frac{\text{closed } p\text{-forms on } \mathcal{U}}{\text{exact } p\text{-forms on } \mathcal{U}}.$$

Example 3.2.7. *Assume $\mathcal{U}$ is connected. Consider $H^0_{\mathrm{dR}}(\mathcal{U})$. A closed 0-form is given by a function f on $\mathcal{U}$ such that*

$$df = 0.$$

So f is a constant function. Therefore

$$H^0_{\mathrm{dR}}(\mathcal{U}) = \mathbb{R}.$$

Example 3.2.8. *Consider* $\mathcal{U} = \mathbb{R}^2 - \{0\}$. *Let*

$$\alpha = \frac{xdy - ydx}{x^2 + y^2}.$$

We can check that $d\alpha = 0$, *hence* α *is a closed* 1*-form. However,* α *is not exact (**Exercise**. Hint: Example 3.1.29). In this case we find* $H^1_{\mathrm{dR}}(\mathcal{U}) \neq 0$.

Definition 3.2.9. $\mathcal{U} \subset \mathbb{R}^n$ is called **star-sharped** if there is a point $x_0 \in \mathcal{U}$ such that for any $x \in \mathcal{U}$, the straight interval connecting x and x_0 is contained in $\mathcal{U}$.

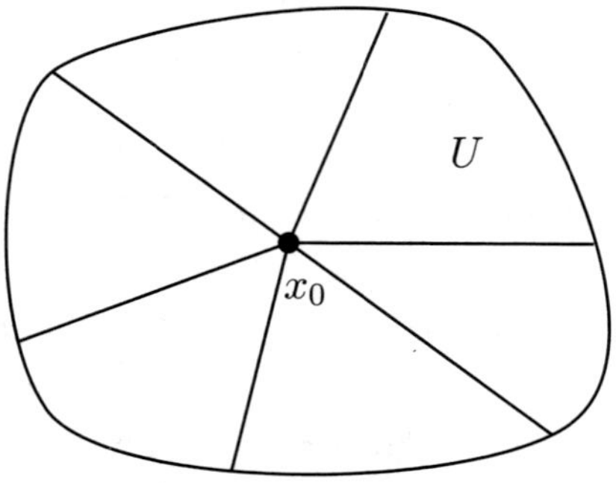

Theorem 3.2.10 (Poincaré Lemma). *If an open subset* $\mathcal{U} \in \mathbb{R}^n$ *is star-shaped, then*

$$H^p_{\mathrm{dR}}(\mathcal{U}) = 0 \qquad \text{for any } p > 0.$$

That is, any closed p-form is exact if $p > 0$.

Proof. Assume $p > 0$, $\alpha \in \Omega^p(\mathcal{U})$ and $d\alpha = 0$. Assume $\mathcal{U}$ is star-sharped with $x_0 = 0$. Consider

$$\begin{aligned} \varphi_t : \mathcal{U} &\longrightarrow \mathcal{U}, \qquad t \leq 0 \\ x &\longmapsto e^t x. \end{aligned}$$

This is well-defined since $\mathcal{U}$ is star-sharped. Note that φ_t is the flow for the vector field

$$V = \sum_i x^i \frac{\partial}{\partial x^i}, \qquad V^i(x) = x^i.$$

In fact, $\frac{\partial}{\partial t}\varphi^i_t(x) = \frac{\partial}{\partial t}(e^t x^i) = e^t x^i = V^i(\varphi_t(x))$. Also

$$\varphi_0 : \mathcal{U} \longrightarrow \mathcal{U}$$

is the identity map, and

$$\varphi_{-\infty} : \mathcal{U} \longrightarrow \{0\} \subset \mathcal{U}$$

is the constant map to the origin. So

$$\varphi^*_0(\alpha) = \alpha, \qquad \varphi^*_{-\infty}(\alpha) = 0.$$

Then

$$\begin{aligned}\alpha &= \varphi_0^*(\alpha) - \varphi_{-\infty}^*(\alpha) \\ &= \int_{-\infty}^0 \frac{d}{dt}(\varphi_t^*(\alpha))dt \\ &= \int_{-\infty}^0 \mathscr{L}_V(\varphi_t^*(\alpha))dt \\ &= \int_{-\infty}^0 (d\iota_V\varphi_t^*\alpha + \iota_V d\varphi_t^*\alpha)dt.\end{aligned}$$

Since $d\varphi_t^*\alpha = \varphi_t^*(d\alpha) = 0$, this is equal to

$$d\int_{-\infty}^0 \iota_V\varphi_t^*\alpha dt = d\beta,$$

where $\beta = \int_{-\infty}^0 \iota_V\varphi_t^*\alpha dt$. Let $\lambda = e^t$ and

$$\phi_\lambda(x) = \lambda x = \varphi_t(x).$$

Then

$$\beta = \int_0^1 \iota_V\phi_\lambda^*(\alpha)\frac{d\lambda}{\lambda}.$$

□

As an example to illustrate the above computation, assume $\alpha = \sum_i \alpha_i(x)dx^i$ is a closed

$$d\alpha = 0.$$

This is equivalent to

$$\partial_j\alpha_i = \partial_i\alpha_j, \qquad \forall i, j.$$

Then

$$\begin{aligned}\beta(x) &= \int_0^1 \iota_V\phi_\lambda^*\left(\sum_i \alpha_i(x)dx^i\right)\frac{d\lambda}{\lambda} \\ &= \int_0^1 \iota_V\left(\sum_i \alpha_i(\lambda x)\lambda dx^i\right)\frac{d\lambda}{\lambda} \\ &= \int_0^1 \sum_i x^i\alpha_i(\lambda x)d\lambda.\end{aligned}$$

We can check $\alpha = d\beta$ or $\alpha_i(x) = \partial_i\beta(x)$ directly:

$$\begin{aligned}\frac{\partial}{\partial x^i}\beta(x) &= \frac{\partial}{\partial x^i}\int_0^1 \sum_j x^j\alpha_j(\lambda x)d\lambda \\ &= \int_0^1 \alpha_i(\lambda x)d\lambda + \int_0^1 \sum_j x^j\lambda\partial_i\alpha_j(\lambda x)d\lambda\end{aligned}$$

$$\begin{aligned}
&= \int_0^1 \alpha_i(\lambda x)d\lambda + \int_0^1 \sum_j x^j \lambda \partial_j \alpha_i(\lambda x) d\lambda \\
&= \int_0^1 \alpha_i(\lambda x)d\lambda + \int_0^1 \lambda \frac{\partial}{\partial \lambda}(\alpha_i(\lambda x))d\lambda \\
&= \int_0^1 \frac{\partial}{\partial \lambda}(\lambda \alpha_i(\lambda x))d\lambda \\
&= \lambda \alpha_i(\lambda x)\Big|_0^1 = \alpha_i(x),
\end{aligned}$$

as required.

Example 3.2.11. *Recall that on the phase space* $\mathbb{R}^{2n}$*, a (time-independent) canonical transformation is a map* $\varphi : (q_i, p_i) \mapsto (Q_i, P_i)$ *such that*

$$\sum_i dp_i \wedge dq_i = \sum_i dP_i \wedge dQ_i.$$

Then

$$d\left(\sum_i p_i dq_i - \sum_i P_i dQ_i\right) = 0.$$

Since $\mathbb{R}^{2n}$ *is obviously star-sharped, there exists a function* F *on* $\mathbb{R}^{2n}$ *such that*

$$\sum_i p_i dq_i - \sum_i P_i dQ_i = dF.$$

F *gives the generating function as we discussed before.*

3.3 Symplectic Form

3.3.1 Symplectic Vector Space

Let V be a finite dimensional real vector space. Let

$$\omega : V \times V \longrightarrow \mathbb{R}$$

be a bilinear map. ω is called skew-symmetric if

$$\omega(u, v) = -\omega(v, u), \qquad \forall u, v \in V.$$

Equivalently, ω defines a linear map

$$\omega : \bigwedge^2 V \longrightarrow \mathbb{R}.$$

We denote the **null space** of ω by

$$N = \{u \in V \,|\, \omega(u, v) = 0, \ \forall v \in V\}.$$

Definition 3.3.1. A skew-symmetric bilinear map ω on V is called **non-degenerate** or **symplectic** if its null space is trivial: $N = \{0\}$. In this case, we say ω is a **symplectic pairing** on V. The pair (V, ω) will be called a **symplectic vector space**.

Proposition 3.3.2. *Let ω be a symplectic pairing on V. Then there exists a basis $\{e_1, e_2, \cdots, e_n, f_1, \cdots, f_n\}$ of V such that*

$$\begin{cases} \omega(e_i, e_j) = \omega(f_i, f_j) = 0 \\ \omega(e_i, f_j) = \delta_{ij} = -\omega(f_j, e_i) \end{cases} \quad \forall i, j.$$

In particular, $\dim V = 2n$ *has to be even.*

Proof. Take a nonzero element $e_1 \in V$. By non-degeneracy, there exists a $f_1 \in V$ such that $\omega(e_1, f_1) = 1$. Let

$$U = \mathrm{Span}_{\mathbb{R}}\{e_1, f_1\}, \qquad U^{\perp} = \{u \in V | \omega(e_1, u) = \omega(f_1, u) = 0\}.$$

Then it is straight forward to check that
① $U \cap U^{\perp} = \{0\}$ and $V = U \oplus U^{\perp}$.
② $\omega|_{U^{\perp}}$ defines a symplectic pairing on $U^{\perp}$.
Then we can continue this process. The proposition follows. □

Remark. Such basis is not unique.

Remark. Let $c_1, c_2, \cdots, c_m$ be an arbitrary basis of V. A skew-symmetric bilinear map ω is represented by a skew-symmetric matrix $\omega_{ij} = \omega(c_i, c_j)$, $\omega_{ij} = -\omega_{ji}$. ω is non-degenerate if (ω_{ij}) is invertible, i.e.,

$$\det(\omega_{ij}) \neq 0.$$

The above proposition says that if A is a skew-symmetric invertible matrix, then there exists an invertible matrix B such that

$$BAB^T = \begin{pmatrix} 0 & I_n \\ -I_n & 0 \end{pmatrix}.$$

Let ω be a skew-symmetric bilinear map on V. It induces a linear map

$$\begin{aligned} \tilde{\omega} : V &\longrightarrow V^* \\ v &\longmapsto \omega(v, -) \end{aligned}$$

The null space is $N = \mathrm{Ker}\, \tilde{\omega}$. In particular, when ω is non-degenerate, $\tilde{\omega}$ is an isomorphism.

Definition 3.3.3. Let (V, ω) be a symplectic vector space. A linear subspace $L \subset V$ is called **isotropic** if $\omega|_{L \times L} = 0$, i.e.,

$$\omega(u, v) = 0, \qquad \forall u, v \in L.$$

Proposition 3.3.4. *Let L be an isotropic subspace of a symplectic vector space (V, ω). Then*

$$\dim L \le \frac{1}{2} \dim V.$$

Proof. The map $\tilde{\omega} : V \to V^*$ induces a map

$$\begin{aligned} \tilde{\omega}_L : L &\longrightarrow (V/L)^* \\ u &\longmapsto \omega(u, -) \end{aligned}$$

$\tilde{\omega}_L$ is injective. In fact, if $u \in L$ and $\tilde{\omega}_L(u) = 0$, then

$$\omega(u, v) = 0, \qquad \forall v \in V \qquad \Longrightarrow \quad u = 0.$$

It follows that $\dim L \le \dim(V/L)$, i.e., $\dim L \le \frac{1}{2} \dim V$. □

Definition 3.3.5. A linear subspace L of a symplectic vector space (V, ω) is called a **Lagrangian subspace** if L is isotropic and

$$\dim L = \frac{1}{2} \dim V.$$

Example 3.3.6. *In Prop 3.3.2, for the basis $\{e_1, e_2, \cdots, e_n, f_1, \cdots, f_n\}$ of V with*

$$\begin{cases} \omega(e_i, e_j) = \omega(f_i, f_j) = 0 \\ \omega(e_i, f_j) = \delta_{ij} \end{cases} \quad \forall i, j,$$

$\mathrm{Span}\{e_1, \cdots, e_n\}$ *and* $\mathrm{Span}\{f_1, \cdots, f_n\}$ *are Lagrangian subspaces of (V, ω).*

3.3.2 Symplectic Form

Let $\mathcal{U} \in \mathbb{R}^{2n}$ be an open subset. Let $\omega \in \Omega^2(\mathcal{U})$ be a 2-form. For each point $p \in \mathcal{U}$, $\omega_p \in \bigwedge^2 T_x^*\mathcal{U}$ defines a skew-symmetric bilinear map

$$\omega_p : T_x\mathcal{U} \times T_x\mathcal{U} \longrightarrow \mathbb{R}.$$

Definition 3.3.7. The 2-form ω is called **symplectic** if ω is closed ($d\omega = 0$) and ω_p is non-degenerate for each $p \in \mathcal{U}$.

In terms of coordinates $\{x^1, \cdots, x^{2n}\}$ of $\mathcal{U}$, we write

$$\omega = \frac{1}{2} \sum_{ij} \omega_{ij}(x) dx^i \wedge dx^j.$$

The non-degeneracy says that the matrix (ω_{ij}) is invertible for each $x \in U$. The closeness says

$$\begin{aligned} d\omega &= \frac{1}{2} \sum_{ijk} \partial_k \omega_{ij} dx^k \wedge dx^i \wedge dx^j \\ &= \frac{1}{3!} \sum_{ijk} \underbrace{(\partial_k \omega_{ij} + \partial_i \omega_{jk} + \partial_j \omega_{ki})}_{\text{totally anti-symmetric in } i,j,k} dx^k \wedge dx^i \wedge dx^j. \end{aligned}$$

Then $d\omega = 0$ if and only if

$$\partial_k \omega_{ij} + \partial_i \omega_{jk} + \partial_j \omega_{ki} = 0, \qquad \forall i, j, k.$$

Example 3.3.8. *Consider the phase space* $\mathbb{R}^{2n}$ *with coordinates* $\{p_1, \cdots, p_n, q_1 \cdots, q_n\}$. *Then the 2-form*

$$\omega = \sum_i dp_i \wedge dq_i$$

is symplectic. The matrix ω_{ij} *takes the form*

$$\begin{pmatrix} 0 & I_n \\ -I_n & 0 \end{pmatrix}.$$

Let $\omega \in \Omega^2(\mathcal{U})$ be a symplectic form. It induces a map

$$\begin{aligned} \tilde{\omega} : \mathrm{Vect}(\mathcal{U}) &\longrightarrow \Omega^1(\mathcal{U}) \\ V &\longmapsto \iota_V \omega = \omega(V, -) \end{aligned}$$

which is an isomorphism due to non-degeneracy of ω.

Definition 3.3.9. Let f be a function on $\mathcal{U}$. We define its associated **Hamiltonian vector field** V_f by $V_f = \tilde{\omega}^{-1}(df)$, i.e.,

$$\iota_{V_f} \omega = df.$$

Example 3.3.10. *Let* $\omega = \sum_i dp_i \wedge dq_i$ *and* $f = f(q, p)$.

$$\tilde{\omega} : \begin{cases} \dfrac{\partial}{\partial q_i} \longmapsto -dp_i \\ \dfrac{\partial}{\partial p_i} \longmapsto dq_i \end{cases}$$

Since $df = \sum_i \partial_{q_i} f dq_i + \partial_{p_i} f dp_i$, *we have*

$$\tilde{\omega}^{-1}(df) = \sum_i (\partial_{q_i} f) \frac{\partial}{\partial p_i} - (\partial_{p_i} f) \frac{\partial}{\partial q_i},$$

i.e.,

$$V_f = \sum_i \frac{\partial f}{\partial q_i} \frac{\partial}{\partial p_i} - \frac{\partial f}{\partial p_i} \frac{\partial}{\partial q_i}.$$

Proposition 3.3.11. *The symplectic form* ω *is invariant under the flow* φ_t *generated by a Hamiltonian vector flow* X_f, *i.e.,*

$$\varphi_t^* \omega = \omega.$$

Proof. It is equivalent to show that

$$\frac{\partial}{\partial t}(\varphi_t^* \omega) = 0,$$

or $\mathscr{L}_{V_f}\omega = 0$. Using Cartan formula,

$$\mathscr{L}_{V_f}\omega = (d\iota_{V_f} + \iota_{V_f} d)\omega = d\iota_{V_f}\omega = ddf = 0.$$ □

As a corollary, this immediately leads to Liouville's Theorem:

$$\varphi_t^*\left(\frac{\omega^n}{n!}\right) = \frac{\omega^n}{n!} \quad \text{or} \quad \mathscr{L}_{V_f}\left(\frac{\omega^n}{n!}\right) = 0.$$

Here φ_t is the flow of the Hamiltonian vector field V_f, and $\frac{\omega^n}{n!}$ is a volume form which becomes the standard volume form on the phase space $\mathbb{R}^{2n}$ where $\omega = \sum_i dp_i \wedge dq_i$.

Poisson bracket

Let ω be a symplectic form on $\mathcal{U}$. Given functions $f, g \in C^\infty(\mathcal{U})$, let $V_f, V_g \in \mathrm{Vect}(\mathcal{U})$ denote the corresponding Hamilton vector fields.

Definition 3.3.12. Define the **Poisson bracket**

$$\{-,-\} : C^\infty(\mathcal{U}) \times C^\infty(\mathcal{U}) \longrightarrow C^\infty(\mathcal{U})$$

by

$$\{f, g\} = \iota_{V_f}\iota_{V_g}\omega.$$

Note that $\iota_{V_g}\omega = dg$, we have

$$\{f, g\} = V_f(g).$$

Example 3.3.13. *Let $\omega = \sum_i dp_i \wedge dq_i$. The Hamiltonian vector field of f is*

$$V_f = \sum_i \frac{\partial f}{\partial q_i}\frac{\partial}{\partial p_i} - \frac{\partial f}{\partial p_i}\frac{\partial}{\partial q_i}.$$

So

$$\{f, g\} = \sum_i \frac{\partial f}{\partial q_i}\frac{\partial g}{\partial p_i} - \frac{\partial f}{\partial p_i}\frac{\partial g}{\partial q_i}.$$

This is the standard Poisson bracket as we have seen before.

In general, let

$$\omega = \frac{1}{2}\sum_{ij} \omega_{ij} dx^i \wedge dx^j.$$

It sends

$$\tilde{\omega} : \frac{\partial}{\partial x^i} \longmapsto \sum_j \omega_{ij} dx^j.$$

Then $\tilde{\omega}^{-1}(dx^i) = \sum_i \omega^{ij} \frac{\partial}{\partial x^j}$, where (ω^{ij}) is the inverse matrix of (ω_{ij}):

$$\sum_k \omega^{ik}\omega_{kj} = \delta^i_j.$$

Then

$$V_f = \tilde{\omega}^{-1}(df) = \tilde{\omega}^{-1}\left(\sum_i \frac{\partial f}{\partial x^i} dx^i\right) = \sum_{i,j} \omega^{ij} \frac{\partial f}{\partial x^i}\frac{\partial}{\partial x^j}.$$

We find the following formula for the Poisson bracket

$$\{f,g\} = \sum_{i,j} \omega^{ij} \frac{\partial f}{\partial x^i}\frac{\partial g}{\partial x^j}.$$

Proposition 3.3.14. *The Poisson bracket satisfies the following properties*
① $\{f,g\} = -\{g,f\}$.
② $\{f,gh\} = \{f,g\}h + g\{f,h\}$.
③ *Jacobi identity:* $\{f,\{g,h\}\} + \{g,\{h,f\}\} + \{h,\{f,g\}\} = 0$.

Proof. ① $\{f,g\} = \iota_{V_f}\iota_{V_g}\omega = -\iota_{V_g}\iota_{V_f}\omega = -\{g,f\}$.
② $\{f,gh\} = V_f(gh) = V_f(g)h + gV_f(h) = \{f,g\}h + g\{f,h\}$.
③ This follows from $d\omega = 0$. Exercise. □

The Poisson bracket and Lie bracket of vector fields are closely related.

Proposition 3.3.15. $[V_f, V_g] = V_{\{f,g\}}$.

Proof.

$$\begin{aligned}\iota_{[V_f,V_g]}\omega &= [\mathscr{L}_{V_f}, \iota_{V_g}]\omega = \mathscr{L}_{V_f}\iota_{V_g}\omega \qquad (\text{Since } \mathscr{L}_{V_f}\omega = 0)\\ &= \mathscr{L}_{V_f} dg = d\iota_{V_f} dg\\ &= d\{f,g\}.\end{aligned}$$

□

This proposition says the following. We have a map

$$\begin{aligned} C^\infty &\longrightarrow \mathrm{Vect}(\mathcal{U})\\ f &\longmapsto V_f\end{aligned}$$

Then this is a map from the Lie algebra $(C^\infty(\mathcal{U}), \{-,-\})$ to the Lie algebra $(\mathrm{Vect}(\mathcal{U}), [-,-])$.

3.3.3 Darboux Theorem

We show that locally any symplectic form looks like the standard symplectic form on the phase space $\mathbb{R}^{2n}$ under a change of coordinate.

Theorem 3.3.16 (Darboux Theorem). *Let ω be a symplectic form defined in some open subset of $\mathbb{R}^{2n}$ containing 0. Then there exists coordinate functions $\{q_1, \cdots, q_n, p_1, \cdots, p_n\}$ defined on a neighborhood $\mathcal{U}$ of 0 such that*

$$\omega = \sum_{i=1}^{n} dp_i \wedge dq_i \qquad \text{on } \mathcal{U}.$$

Before we prove the theorem, we need a formula on the flow of time-dependent vector field. Let V_t be a time-dependent vector field and φ_t be the corresponding flow on some region $\mathcal{U}$. Here t is the time variable. The flow equation is

$$\frac{\partial}{\partial t}\varphi_t(x) = V_t(\varphi_t(x))$$

for point x and time t where defined.

Proposition 3.3.17. *Let α be a differential form on $\mathcal{U}$. Then*

$$\frac{\partial}{\partial t}(\varphi_t^*\alpha) = \varphi_t^*(\mathscr{L}_{V_t}\alpha)$$

wherever defined.

Proof. We show the above formula is true for $\alpha = f$ and $\alpha = df$ where f is a function. Then the above formula holds for a general differential form.

- $\alpha = f$.

$$\begin{aligned}\frac{\partial}{\partial t}(\varphi_t^* f)(x) &= \frac{\partial}{\partial t}(f(\varphi_t(x)) = \frac{\partial \varphi_t^i(x)}{\partial t}\frac{\partial f}{\partial x^i}(\varphi_t(x)) = V_t^i(\varphi_t(x))\frac{\partial f}{\partial x^i}(\varphi_t(x)) \\ &= (\varphi_t^*(\mathscr{L}_{V_t} f))(x).\end{aligned}$$

- $\alpha = df$.

$$\frac{\partial}{\partial t}(\varphi_t^* df) = d(\frac{\partial}{\partial t}\varphi_t^* f) = d(\varphi_t^*\mathscr{L}_{V_t} f) = \varphi_t^*(d\mathscr{L}_{V_t} f) = \varphi_t^*\mathscr{L}_{V_t}(df). \qquad \square$$

Proof of Darboux Theorem. The proof is based on the so-called Moser's trick. Let

$$\omega = \frac{1}{2}\sum_{i,j}\omega_{ij}(x)dx^i \wedge dx^j.$$

Let $\omega_0 = \frac{1}{2}\sum_{i,j}\omega_{ij}(0)dx^i \wedge dx^j$. Since $d(\omega - \omega_0) = 0$, Poincaré Lemma implies that $\omega - \omega_0 = d\beta$ in some open neighborhood $\mathcal{U}$ of 0 for a 1-form β. Put

$$\omega_t = \omega_0 + t d\beta.$$

Taking $\mathcal{U}$ smaller if necessary, we may assume ω_t is non-degenerate on $\mathcal{U}$ for any $t \in [0, 1]$. Define a time-dependent vector field V_t on $\mathcal{U}$ by

$$\iota_{V_t}\omega_t = -\beta.$$

Since $(\omega-\omega_0)|_{x=0}=0$, β and hence V_t vanishes at $x=0$. Let φ_t be the flow of V_t. Then $\varphi_t(0)=0$. We can find a smaller neighborhood $\tilde{\mathcal{U}}$ of 0 such that

$$\varphi_t : \tilde{\mathcal{U}} \longrightarrow \mathcal{U}$$

is defined for $t \in [0,1]$. Consider $\varphi_t^*\omega_t$ on $\tilde{\mathcal{U}}$. By the Proposition above,

$$\frac{\partial}{\partial t}(\varphi_t^*\omega_t) = \varphi_t^*(\mathscr{L}_{V_t}\omega_t + \frac{\partial}{\partial t}\omega_t) = \varphi_t^*(d\iota_{V_t}\omega_t + d\beta) = \varphi_t^*(-d\beta + d\beta) = 0.$$

So

$$\varphi_t^*\omega_t = \omega_0 \qquad \text{for all } t \in [0,1].$$

It follows that

$$\varphi_1^*\omega = \omega_0$$

or equivalently $\omega = (\varphi_1^{-1})^*\omega_0$. Let

$$\varphi_1^{-1} : (x^1, \cdots, x^{2n}) \longmapsto (y^1 = y^1(x), \cdots, y^{2n} = y^{2n}(x)).$$

Then the functions $y^1(x)$, $\cdots$, $y^{2n}(x)$ will satisfy

$$\omega = \frac{1}{2}\sum_{i,j}\omega_{ij}(x)dx^i \wedge dx^j = \frac{1}{2}\sum_{i,j}\omega_{ij}(0)dy^i \wedge dy^j.$$

By Prop 3.3.2, a further linear transformation $\{y^1, \cdots, y^{2n}\} \mapsto \{q_1, \cdots, q_n, p_1, \cdots, p_n\}$ gives coordinate functions $q_1(x)$, $\cdots$, $q_n(x)$, $p_1(x)$, $\cdots$, $p_n(x)$ such that $\omega = \sum_{i=1}^n dp_i(x) \wedge dq_i(x)$. □

3.4 Geometry of Canonical Transformations

3.4.1 Canonical Transformation Revisited

Now we consider the standard symplectic form

$$\omega = \sum_{i=1}^n dp_i \wedge dq_i$$

on the phase space $\mathcal{U} \subset \mathbb{R}^{2n}$. We consider the extended phase space $\mathcal{U}\times\mathbb{R}$ parametrized by $\{q_1, \cdots, q_n, p_1, \cdots, p_n, t\}$. Let $\mathscr{H} = \mathscr{H}(q,p,t)$ be a time-dependent (Hamiltonian) function defined on $\mathcal{U}\times\mathbb{R}$. We consider the following 2-form

$$\omega_{\mathscr{H}} = \omega - d\mathscr{H} \wedge dt \qquad \text{on } \mathcal{U}\times\mathbb{R}.$$

Let $\pi : \mathcal{U}\times\mathbb{R} \to \mathbb{R}$ denote the projection to the time.

Proposition 3.4.1. *There exists a unique vector field $\mathcal{X}_{\mathscr{H}}$ on $\mathcal{U}\times\mathbb{R}$ such that $\pi_*(\mathcal{X}_{\mathscr{H}}) = \frac{\partial}{\partial t}$ and $\iota_{\mathcal{X}_{\mathscr{H}}}\omega_{\mathscr{H}} = 0$.*

Proof. Since $\pi_*(\mathcal{X}_{\mathscr{H}}) = \frac{\partial}{\partial t}$, $\mathcal{X}_{\mathscr{H}}$ must be of the form

$$\mathcal{X}_{\mathscr{H}} = \frac{\partial}{\partial t} - \sum_i \left(a^i(q,p,t)\frac{\partial}{\partial q_i} + b^i(q,p,t)\frac{\partial}{\partial p_i} \right)$$

for some functions a^i, b^i on $U \times \mathbb{R}$. Since

$$\omega_{\mathscr{H}} = \sum_i dp_i \wedge dq_i - \sum_i (\partial_{q_i}\mathscr{H} dq_i + \partial_{p_i}\mathscr{H} dp_i) \wedge dt,$$

$$\iota_{\mathcal{X}_{\mathscr{H}}}\omega_{\mathscr{H}} = \sum_i (a^i dp_i - b^i dq_i) + \sum_i (\partial_{q_i}\mathscr{H} dq_i + \partial_{p_i}\mathscr{H} dp_i) + (a^i \partial_{q_i}\mathscr{H} + b^i \partial_{p_i}\mathscr{H})dt = 0,$$

$$\Longrightarrow \qquad a^i = -\partial_{p_i}\mathscr{H}, \qquad b^i = \partial_{q_i}\mathscr{H}.$$

So

$$\mathcal{X}_{\mathscr{H}} = \frac{\partial}{\partial t} - \sum_i \left(\partial_{q_i}\mathscr{H}\frac{\partial}{\partial p_i} - \partial_{p_i}\mathscr{H}\frac{\partial}{\partial q_i} \right)$$ □

We observe that the vector field $\mathcal{X}_{\mathscr{H}}$ is closely related to Hamilton's equations. In fact, let $\gamma(t) : I \to \mathcal{U} \times \mathbb{R}$ be the integral curve of the vector field $\mathcal{X}_{\mathscr{H}}$ with initial condition

$$\gamma(0) = (q_0, p_0, 0).$$

Write $\gamma(t) = (q_i(t), p_i(t), t)$. Then the flow equation $\frac{d}{dt}\gamma(t) = \mathcal{X}_{\mathscr{H}}(\gamma(t))$ reads

$$\begin{cases} \dfrac{dq_i(t)}{dt} = \dfrac{\partial \mathscr{H}}{\partial p_i} \\ \dfrac{dp_i(t)}{dt} = -\dfrac{\partial \mathscr{H}}{\partial q_i} \end{cases}$$

They are precisely Hamilton's Equations. This motivates the following

Definition 3.4.2. Let $\mathcal{U} \subset \mathbb{R}^{2n}$ and $\mathscr{H} = \mathscr{H}(q,p,t)$. A diffeomorphism

$$\varphi : \mathcal{U} \times \mathbb{R} \longrightarrow \mathcal{V} \times \mathbb{R}, \qquad \mathcal{V} \subset \mathbb{R}^{2n}$$

is called a canonical transformation if

① $\pi \circ \varphi = \pi$.

$$\begin{array}{ccc} \mathcal{U} \times \mathbb{R} & \xrightarrow{\quad\varphi\quad} & \mathcal{V} \times \mathbb{R} \\ & {\scriptstyle\pi}\searrow \quad \swarrow{\scriptstyle\pi} & \\ & \mathbb{R} & \end{array}$$

② There exists a function $\mathscr{K}$ on $\mathcal{V} \times \mathbb{R}$ such that

$$\varphi^* \omega_{\mathscr{K}} = \omega_{\mathscr{H}}.$$

The condition ① says that φ is expressed as

$$\varphi : (q_i, p_i, t) \longmapsto (Q_i(q,p,t), P_i(q,p,t), t).$$

Here $\{q_i, p_i\}$ are the coordinates on $\mathcal{U}$ and $\{Q_i, P_i\}$ are the coordinates on $\mathcal{V}$.

The condition ② says

$$\sum_i dp_i \wedge dq_i - d\mathscr{H} \wedge dt = \sum_i dP_i \wedge dQ_i - d\mathscr{K} \wedge dt,$$

$$\Longrightarrow \quad d\left(\left(\sum_i p_i dq_i - \mathscr{H} dt\right) - \left(\sum_i P_i dQ_i - \mathscr{K} dt\right)\right) = 0.$$

Assume $\mathcal{U}$ is a star-sharped region, then Poincaré Lemma implies

$$\left(\sum_i p_i dq_i - \mathscr{H} dt\right) - \left(\sum_i P_i dQ_i - \mathscr{K} dt\right) = dF$$

for some function F on $\mathcal{U} \times \mathbb{R}$. This is the generating function of the canonical transformation that we have discussed before. Moreover, $\varphi^* \omega_{\mathscr{K}} = \omega_{\mathscr{H}}$ implies that

$$\varphi_*(\mathcal{X}_{\mathscr{H}}) = \mathcal{X}_{\mathscr{K}},$$

and so φ maps integral curve of $\mathcal{X}_{\mathscr{H}}$ to that of $\mathcal{X}_{\mathscr{K}}$. In other words, φ maps Hamilton's equations for $\mathscr{H}$ to Hamilton's equations for $\mathscr{K}$. This gives a geometric reasoning of Prop 2.4.5 on why canonical transformations keep the form of Hamiltonian dynamics.

3.4.2 Poincaré's Integral Invariant

Consider the extended phase space $\mathcal{U} \times \mathbb{R}$. Let $\gamma \subset \mathcal{U}$ be a closed curve in U. Let $\tilde{\gamma}$ be the evolution of γ under some time. See the picture below.

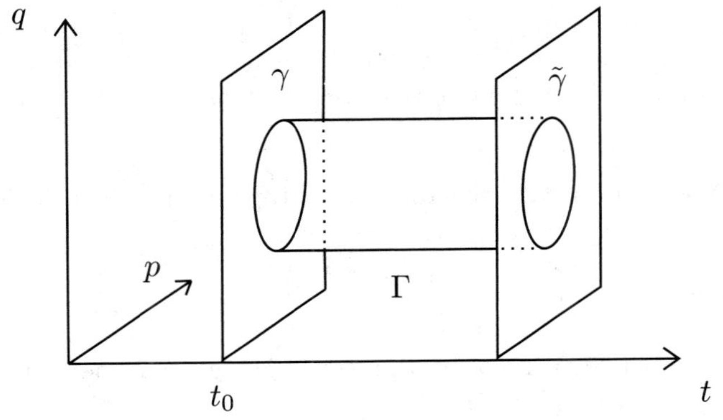

In the extended phase space, γ and $\tilde{\gamma}$ are bounded by a tube Γ, whose horizontal curves are integral curves of $\mathcal{X}_{\mathscr{H}}$. Since $\iota_{\mathcal{X}_{\mathscr{H}}}\omega_{\mathscr{H}} = 0$, we have

$$\int_\Gamma \omega_{\mathscr{H}} = 0.$$

On the other hand, we have $\omega_{\mathscr{H}} = d\alpha$, where

$$\alpha = \sum_i p_i dq_i - \mathscr{H} dt.$$

By Stokes Theorem, we find

$$\oint_{\tilde{\gamma}} \alpha - \oint_\gamma \alpha = \int_{\partial\Gamma} \alpha = \int_\Gamma d\alpha = \int_\Gamma \omega = 0$$

$$\implies \quad \oint_\gamma \alpha = \oint_{\tilde{\gamma}} \alpha.$$

Since γ and $\tilde{\gamma}$ lie on the slice with $t = \text{const}$, we find

$$\oint_\gamma \left(\sum_i p_i dq_i\right) = \oint_{\tilde{\gamma}} \left(\sum_i p_i dq_i\right).$$

i.e., $\oint_\gamma (\sum_i p_i dq_i)$ is a constant of motion. We will come back to this in our later discussion of integrable systems.

3.5 Symplectic Manifold

3.5.1 Smooth Manifold

Coordinate chart

We generalize our discussion so far to a class of spaces called **differentiable manifolds**, which in general can not be considered as open subsets in $\mathbb{R}^n$. Roughly speaking, an n-dimensional manifold is a topological space M which locally looks like an open subset of $\mathbb{R}^n$.

Definition 3.5.1. A **coordinate chart** for a topological space M is a pair $(\mathcal{U}, \varphi)$ where

- $\mathcal{U}$ is an open subset of M.
- $\varphi : \mathcal{U} \to \mathbb{R}^n$ defines a homeomorphism of $\mathcal{U}$ onto an open subset of $\mathbb{R}^n$, i.e., $\varphi(\mathcal{U}) \subset \mathbb{R}^n$ is open and $\varphi : \mathcal{U} \to \varphi(\mathcal{U})$ is a homeomorphism.

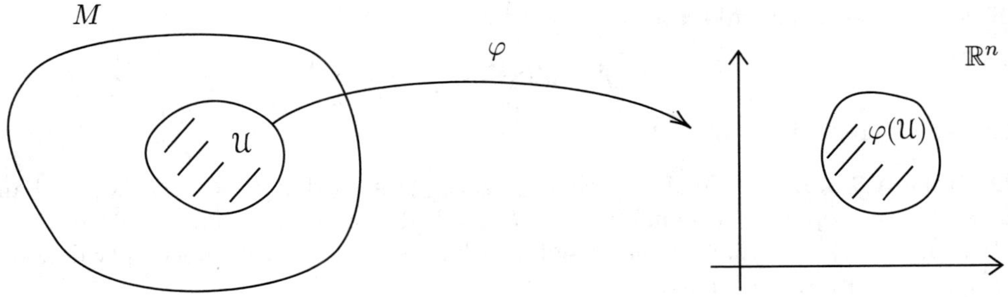

Definition 3.5.2. A **topological manifold** is a Hausdorff space M with a countable basis for its topology such that every point of M lies in some open subset of a coordinate chart.

In other words, M can be described locally by an open subset of $\mathbb{R}^n$ via some coordinate chart. We say M is an n-dimensional topological manifold.

Consider now two charts $(\mathcal{U}_1, \varphi_1)$ and $(\mathcal{U}_2, \varphi_2)$ with $\mathcal{U}_1 \cap \mathcal{U}_2 \neq \varnothing$.

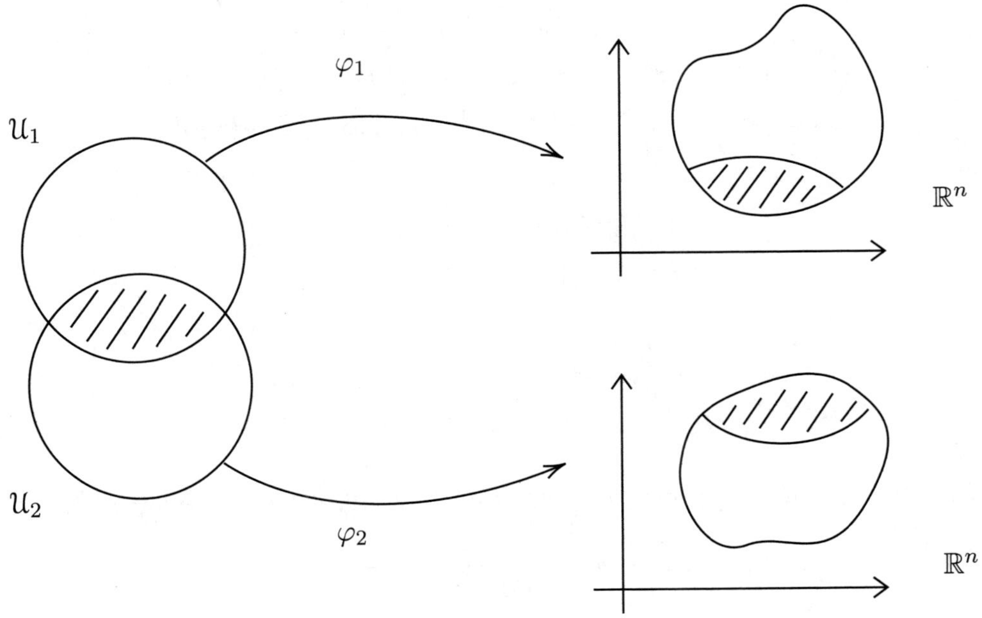

Let $\{x^1, \cdots, x^n\}$ be the coordinate chart on $\varphi_1(\mathcal{U}_1) \subset \mathbb{R}^n$, $\{y^1, \cdots, y^n\}$ be the coordinate chart on $\varphi_2(\mathcal{U}_2) \subset \mathbb{R}^n$. Then on the intersection, we have a homeomorphism of open subsets of $\mathbb{R}^n$

$$\begin{aligned} \varphi_2 \circ \varphi_1^{-1} : \varphi_1(\mathcal{U}_1 \cap \mathcal{U}_2) &\longrightarrow \varphi_2(\mathcal{U}_1 \cap \mathcal{U}_2) \\ (x^1, \cdots, x^n) &\longmapsto (y^1, \cdots, y^n) \end{aligned}$$

This describes a change of coordinates

$$y^i = y^i(x^1, \cdots, x^n)$$

for two different charts of M.

Definition 3.5.3. A collection of charts $\{(\mathcal{U}_r, \varphi_r)\}$ such that $M = \cup_r \mathcal{U}_r$ is called an **atlas** for the topological manifold M. M is called a smooth manifold if there is an atlas $\{(\mathcal{U}_r, \varphi_r)\}$ such that all changes of coordinates $\varphi_s \circ \varphi_r^{-1}$ with nonempty domain of definition are smooth maps.

We will mainly consider smooth manifolds.

Example 3.5.4. *Let $\mathcal{U} \subset \mathbb{R}^n$ be an open subset. Then $\mathcal{U}$ is a smooth manifold with an atlas consisting of just one chart $(\mathcal{U}, i : \mathcal{U} \to \mathbb{R}^n)$.*

Example 3.5.5 (n-dim sphere). *Let*

$$S^n = \left\{ (x^1, x^2, \cdots, x^{n+1}) \in \mathbb{R}^{n+1} \,\middle|\, \sum_{i=1}^{n+1} (x^i)^2 = 1 \right\}.$$

Put

$$\mathcal{U}_1 = \{(x^1, \cdots, x^{n+1}) \in S^n \mid x^{n+1} > -1\} = S^n - \{\textit{south pole}\},$$
$$\mathcal{U}_2 = \{(x^1, \cdots, x^{n+1}) \in S^n \mid x^{n+1} < 1\} = S^n - \{\textit{north pole}\},$$

and

$$\varphi_1 : \mathcal{U}_1 \longrightarrow \mathbb{R}^n$$
$$(x^1, \cdots, x^{n+1}) \longmapsto \frac{1}{1 + x^{n+1}}(x^1, \cdots, x^n)$$

$$\varphi_2 : \mathcal{U}_2 \longrightarrow \mathbb{R}^n$$
$$(x^1, \cdots, x^{n+1}) \longmapsto \frac{1}{1 - x^{n+1}}(x^1, \cdots, x^n)$$

Then $\{(\mathcal{U}_1, \varphi_1), (\mathcal{U}_2, \varphi_2)\}$ gives an atlas. The change of coordinates is given by

$$\varphi_1 \circ \varphi_2^{-1}(y^1, \cdots, y^n) = \frac{1}{(y^1)^2 + \cdots + (y^n)^2}(y^1, \cdots, y^n)$$

which is smooth. So S^n is a smooth manifold.

Example 3.5.6 (Complex projective space $\mathbb{C}P^n$). *Consider*

$$\mathbb{C}P^n = \left(\mathbb{C}^{n+1} - \{0\}\right) / \mathbb{C}^*.$$

Here the quotient relation is

$$(z^0, z^1, \cdots, z^n) \sim (\lambda z^0, \lambda z^1, \cdots, \lambda z^n) \qquad \textit{for } \lambda \in \mathbb{C}^*.$$

The corresponding equivalence class will be denoted by $[z^0, z^1, \cdots, z^n]$. $\mathbb{C}P^n$ *can be covered by the following open subsets*

$$\mathcal{U}_0 = \{z^0 \neq 0\}, \qquad \mathcal{U}_1 = \{z^1 \neq 0\}, \qquad \cdots \qquad ,\mathcal{U}_n = \{z^n \neq 0\}.$$

These become an atlas by associating chart maps

$$\varphi_i : \mathcal{U}_i \longrightarrow \mathbb{C}^n \simeq \mathbb{R}^{2n}$$
$$[z^0, \cdots, z^n] \longmapsto \left(\frac{z^0}{z^i}, \frac{z^1}{z^i}, \cdots, \widehat{\frac{z^i}{z^i}}, \cdots, \frac{z^n}{z^i}\right).$$

The change of coordinates $\varphi_j \circ \varphi_i^{-1}$ *are all smooth, so* $\mathbb{C}P^n$ *is a smooth manifold. In fact, functions* $\varphi_j \circ \varphi_i^{-1}$ *are all holomorphic maps, and this is an example of complex manifolds.*

Smooth functions

Definition 3.5.7. Let M be a smooth manifold. A function

$$f : M \longrightarrow \mathbb{R}$$

is called smooth at $x \in M$ if there is a chart $(\mathcal{U}, \varphi)$ with $x \in \mathcal{U}$ such that

$$f \circ \varphi^{-1} : \underset{\mathbb{R}^n}{\underset{\cap}{\varphi(\mathcal{U})}} \longrightarrow \mathbb{R}$$

is smooth at $\varphi(x)$.

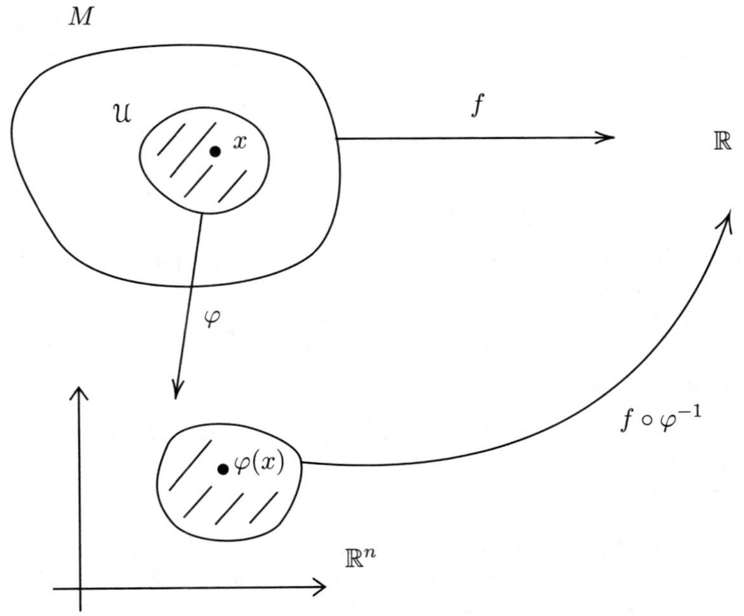

This property is independent of the choice of chart for the smooth structure of M. In fact, let $(\tilde{\mathcal{U}}, \tilde{\varphi})$ be another chart with $x \in \tilde{\mathcal{U}}$, then we have

$$f \circ \tilde{\varphi}^{-1} = (\varphi \circ \tilde{\varphi}^{-1})^* (f \circ \varphi^{-1}) = (f \circ \varphi^{-1}) \circ (\varphi \circ \tilde{\varphi}^{-1}).$$

Since $\varphi \circ \tilde{\varphi}^{-1}$ is smooth, $f \circ \varphi^{-1}$ is smooth at $\varphi(x)$ if and only if $f \circ \tilde{\varphi}^{-1}$ is smooth at $\tilde{\varphi}(x)$.

Definition 3.5.8. A function $f : M \to \mathbb{R}$ is smooth if f is smooth at each point of M.

Differential forms

Differential forms behave like functions and can be defined in a similar way on manifolds as that of functions.

Definition 3.5.9. A (smooth) differential p-form on M is an assignment of a p-form $\alpha_\varphi \in \Omega^p(\varphi(\mathcal{U}))$ for each chart $\varphi : \mathcal{U} \to \mathbb{R}^n$ such that for any two charts $(\mathcal{U}_1, \varphi_1)$ and $(\mathcal{U}_2, \varphi_2)$ with nonempty intersection, we have $\alpha_{\varphi_2} = (\varphi_1 \circ \varphi_2^{-1})^* \alpha_{\varphi_1} \in \Omega^p(\varphi_2(\mathcal{U}_1 \cap \mathcal{U}_2))$.

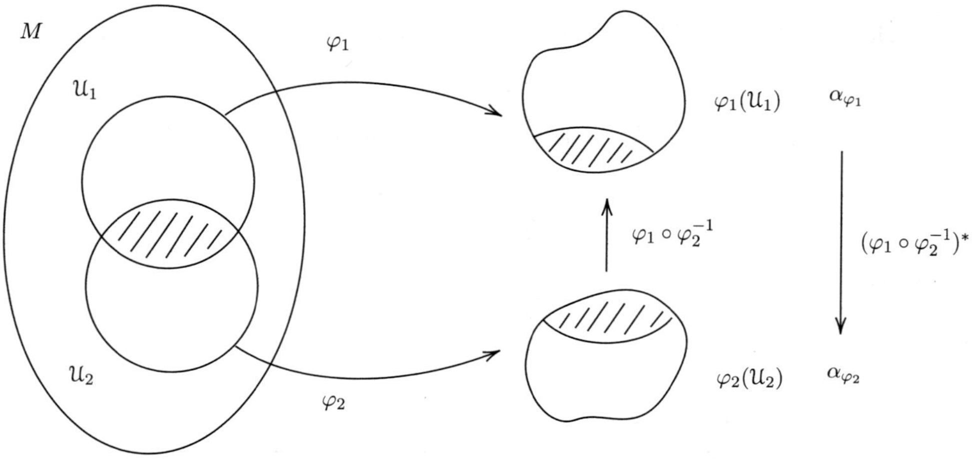

We will denote

$$\Omega^p(M) = \{\text{smooth } p\text{-forms on } M\}.$$

The exterior derivative and wedge product are all compatible with pull-backs, hence we have well-defined operators

$$d : \Omega^p(M) \longrightarrow \Omega^{p+1}(M), \qquad \wedge : \Omega^p(M) \times \Omega^q(M) \longrightarrow \Omega^{p+q}(M).$$

They satisfy the Leibnitz rule

$$d(\alpha \wedge \beta) = d\alpha \wedge \beta + (-1)^\alpha \alpha \wedge d\beta, \qquad \alpha \in \Omega^p(M).$$

Vector fields

Vector fields can be push-forward under coordinate transformations. This allows us to define vector fields on manifolds via local descriptions on charts such that they are compatible under coordinate transformations.

Definition 3.5.10. A (smooth) vector field V on M is an assignment of a (smooth) vector field $V_\varphi \in \mathrm{Vect}(\varphi(\mathcal{U}))$ for each chart $\varphi : \mathcal{U} \to \mathbb{R}^n$ such that for any two charts $(\mathcal{U}_1, \varphi_1)$ and $(\mathcal{U}_2, \varphi_2)$ with nonempty intersection, we have $V_{\varphi_1} = (\varphi_1 \circ \varphi_2^{-1})_* V_{\varphi_2} \in \mathrm{Vect}(\varphi_1(\mathcal{U}_1 \cap \mathcal{U}_2))$.

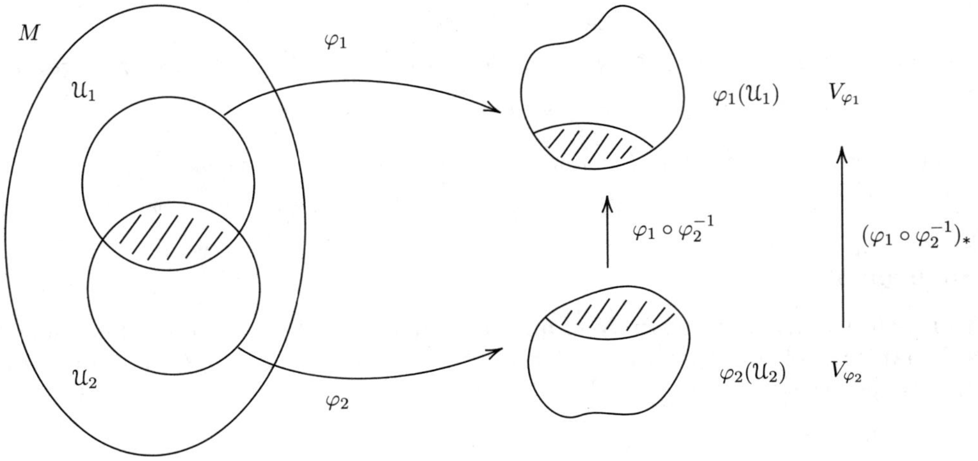

We will denote

$$\mathrm{Vect}(M) = \{\text{smooth vector fields on } M\}\,.$$

We can apply similar idea to define a tangent vector at a point $p \in M$. It is locally described by a tangent vector in some chart, and compatible under coordinate transformations. Let

$$T_pM = \{\text{tangent vectors at } p \in M\}\,.$$

Example 3.5.11. *Let $\gamma : (-\varepsilon, \varepsilon) \to M$ be a path on M and $\gamma(0) = p$. Then we have a well-defined element*

$$\gamma'(0) \in T_pM$$

as before.

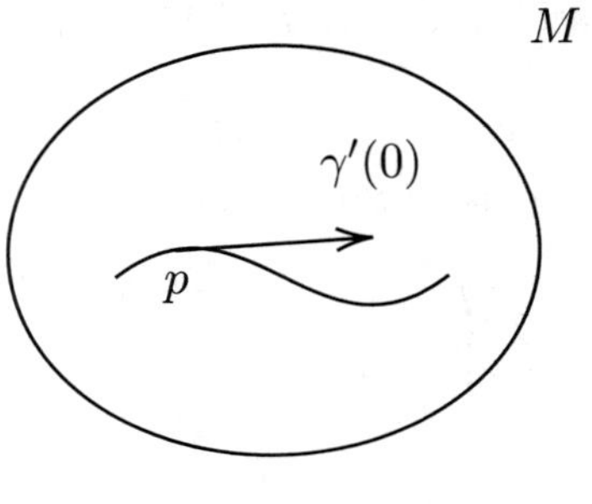

In fact, let $p \in \mathcal{U}$, and

$$\varphi_1 : \mathcal{U}_1 \longrightarrow \mathbb{R}^n$$

be a chart with coordinates $x^1, \cdots, x^n$. The path is expressed in the chart by

$$\varphi_1 \circ \gamma(t) = (x^1(t), \cdots, x^n(t)).$$

Then the tangent vector $\gamma'(0)$ is expressed in this chart by

$$(\varphi_1 \circ \gamma)'(0) \in T_{\varphi_1(p)}\mathbb{R}^n.$$

In another chart $\varphi_2 : \mathcal{U}_2 \to \mathbb{R}^n$ with coordinates $(y^1, \cdots, y^n)$, $\gamma'(0)$ is represented by

$$(\varphi_2 \circ \gamma)'(0) \in T_{\varphi_2(p)}\mathbb{R}^n.$$

They are compatible by push-forward under coordinate transformation

$$(\varphi_2 \circ \gamma)'(0) = (\varphi_2 \circ \varphi_1^{-1})_*(\varphi_1 \circ \gamma)'(0).$$

Submanifold

Definition 3.5.12. A subset N of an n-dim manifold M is called a p-dimensional **submanifold** of M, if for each point $x_0 \in N$, there is a chart $(\mathcal{U}, \varphi)$ for M with $x_0 \in \mathcal{U}$ such that

$$\varphi(\mathcal{U} \cap N) = \{(x^1, \cdots, x^p, x^{p+1}, \cdots, x^n) \in \varphi(\mathcal{U}) \mid x^{p+1} = \cdots = x^n = 0\}.$$

Then the collection of the corresponding

$$\begin{aligned} \bar{\varphi} : \mathcal{U} \cap N &\longrightarrow \mathbb{R}^p \\ x &\longmapsto (x^1, \cdots, x^p) \end{aligned}$$

gives an atlas for N.

Example 3.5.13. *An open subset of M is a submanifold.*

Example 3.5.14. *Let*

$$\begin{aligned} f : X &\longrightarrow \mathbb{R}^m \\ x &\longmapsto (f_1(x), \cdots, f_m(x)) \end{aligned}$$

be a smooth map. Let

$$N = f^{-1}(0) = \{x \in X \mid f_1(x) = \cdots = f_m(x) = 0\}.$$

Assume f has rank m at each $x \in N$. This means that in some chat $(\mathcal{U}, \varphi)$ with $x \in \mathcal{U}$, the map

$$g = f \circ \varphi^{-1} : \varphi(\mathcal{U}) \longrightarrow \mathbb{R}^m$$

has rank m at $\varphi(x)$, i.e., the rank of this matrix

$$\begin{pmatrix} \frac{\partial g_1}{\partial x^1} & \cdots & \frac{\partial g_1}{\partial x^n} \\ \vdots & & \vdots \\ \frac{\partial g_m}{\partial x^1} & \cdots & \frac{\partial g_m}{\partial x^n} \end{pmatrix}$$

equals m at the point $\varphi(x)$. Equivalently, the push-forward

$$g_* : T_{\varphi(x)}\varphi(\mathcal{U}) \longrightarrow T_{f(x)}\mathbb{R}^m \qquad \textit{is surjective.}$$

This property is independent of the choice of charts. Then $N \subset X$ is a smooth submanifold of dimension $n-m$. This essentially follows from the Implicit Function Theorem.

For example, consider

$$\begin{aligned} f : \mathbb{R}^{n+1} &\longrightarrow \mathbb{R} \\ (x^0, \cdots, x^n) &\longmapsto \sum_{i=0}^{n} (x^i)^2 - 1. \end{aligned}$$

Then $S^n = f^{-1}(0)$, which is the n-sphere, is a submanifold of $\mathbb{R}^{n+1}$.

Example 3.5.15. *The previous example can be further generalized to a smooth map*

$$f : X \longrightarrow M$$

between two smooth manifolds where $\dim X = n$, $\dim M = m$. Let $m_0 \in M$ and

$$N = f^{-1}(m_0) \subset X.$$

Assume f has rank m at each point $x \in N$. This means that in some chart $(\mathcal{U}, \varphi)$ of X with $x \in \mathcal{U}$ and some chart $(\mathcal{V}, \phi)$ of M with $m_0 \in \mathcal{V}$, and such that $\varphi(U) \subset \mathcal{V}$, the map

$$\phi \circ f \circ \varphi^{-1} : \varphi(\mathcal{U}) \longrightarrow \phi(\mathcal{V})$$

has rank m at $\varphi(x)$. This again is independent of the choice of charts. Then $N \subset X$ is a submanifold of dimension $n - m$.

Example 3.5.16. *Consider the map*

$$\begin{aligned} \pi : S^3 = \{(z_1, z_2) \in \mathbb{C}^2 \,|\, |z_1|^2 + |z_2|^2 = 1\} &\longrightarrow S^2 \simeq \mathbb{C}P^1 \\ (z_1, z_2) &\longmapsto [z_1, z_2] \end{aligned}$$

π is a smooth map. For each point $p \in S^2$, we have $\pi^{-1}(p) = S^1$. For example, consider the north pole $n = [1, 0]$, then

$$\pi^{-1}(n) = \{(z, 0) \,|\, |z|^2 = 1\}.$$

*The map π leads to a family of submanifolds of S^3, each of which is diffeomorphic to S^1. This is the famous **Hopf fibration**.*

Cartan calculus

We still have the following well-defined operators on manifolds

- interior product:

$$\begin{aligned} \iota : \mathrm{Vect}(M) \times \Omega^p(M) &\longrightarrow \Omega^{p-1}(M) \\ (V, \alpha) &\longmapsto \iota_V \alpha \end{aligned}$$

- Lie derivative:

$$\begin{aligned}\mathscr{L} : \mathrm{Vect}(M) \times \Omega^p(M) &\longrightarrow \Omega^p(M)\\ (V, \alpha) &\longmapsto \mathscr{L}_V \alpha\end{aligned}$$

They are described in local coordinate charts as the same formula before, and compatible with coordinate transformations between differential charts (check it!). The following theorem still holds from its local version.

Theorem 3.5.17. *Let $V, W \in \mathrm{Vect}(M)$. Then as operators on $\Omega^\bullet(M)$,*
① $[\mathscr{L}_V, \mathscr{L}_W] := \mathscr{L}_V \mathscr{L}_W - \mathscr{L}_W \mathscr{L}_V = \mathscr{L}_{[V,W]}$.
② $\iota_V \iota_W + \iota_W \iota_V = 0$.
③ $[\mathscr{L}_V, d] := \mathscr{L}_V d - d\mathscr{L}_V = 0$.
④ $\iota_V d + d\iota_V = \mathscr{L}_V$.
⑤ $[\mathscr{L}_V, \iota_W] := \mathscr{L}_V \iota_W - \iota_W \mathscr{L}_V = \iota_{[V,W]}$.

3.5.2 Symplectic Manifold

Definition 3.5.18. A symplectic manifold is a pair (M, ω) where M is a smooth manifold, ω is a smooth 2-form on M, such that ω is a symplectic form in any local chart. Equivalently, $d\omega = 0$ and $\omega|_p : T_pM \times T_pM \to \mathbb{R}$ is non-degenerate for each $p \in M$.

In particular, (T_pM, ω_p) is a symplectic vector space for each $p \in M$.

Let (M, ω) be a symplectic manifold. Let f be a smooth function on M. Then f defines a Hamiltonian vector field $V_f \in \mathrm{Vect}(M)$ by the equation

$$\iota_{V_f} \omega = df.$$

In particular, $\mathscr{L}_{V_f} \omega = 0$ still holds.

Similarly, we can define Poisson bracket for two functions f, g by

$$\{f, g\} = \iota_{V_f} \iota_{V_g} \omega = V_f(g).$$

$\{-, -\}$ is skew-symmetric, satisfies Leibnitz rule, and satisfies Jacobi identity. They all follows from the local formula that we have established before.

Example 3.5.19. *Let $M = S^2 \subset \mathbb{R}^3$. Let $p \in S^2$. Then T_pS^2 can be identified with vectors in $\mathbb{R}^3$ orthogonal to p.*

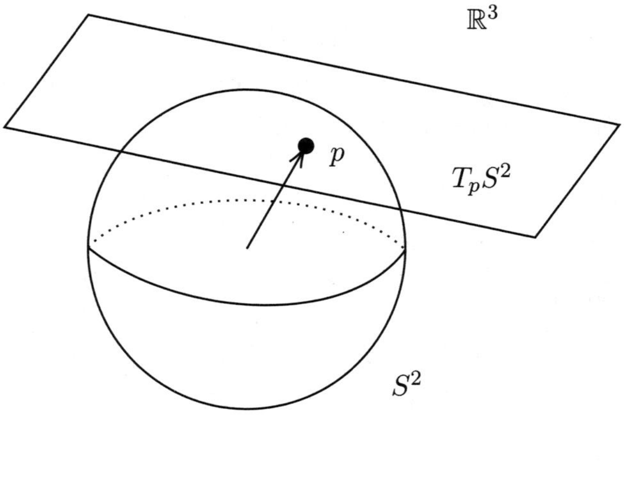

The standard symplectic form ω on S^2 is induced by the symplectic pairing on each T_pS^2:

$$\omega_p(u,v) = \langle p, u \times v \rangle, \qquad u, v \in T_pS^2.$$

This is non-degenerate (check it!). The form ω is closed since it is a top form (dim $S^2 = 2$), so $d\omega = 0$. It is a good exercise to write ω explicitly in local coordinates.

3.5.3 Lagrangian Submanifold

There is a very important class of submanifolds on symplectic manifolds as follows.

Definition 3.5.20. A submanifold $j : L \subset M$ of a symplectic manifold (M, ω) is called a **Lagrangian submanifold** if

$$\omega|_L := j^*\omega = 0, \qquad \text{and} \qquad \dim L = \frac{1}{2}\dim M.$$

This is equivalent to saying that for each point $p \in L$, the tangent space T_pL is a linear Lagrangian subspace (see Definition 3.3.5) of T_pM.

Example 3.5.21. *Consider the symplectic manifold $(\mathbb{R}^{2n}, \omega)$ with $\omega = \sum_i dp_i \wedge dq_i$. Given n functions $f_1(q), \cdots, f_n(q)$ on q_i's, consider the n-dim submanifold defined by*

$$L_f := \{(q_i, p_i) \in \mathbb{R}^{2n} \,|\, p_i = f_i(q) \ i = 1, \cdots, n\}.$$

This submanifold L_f is diffeomorphic to $\mathbb{R}^n$ and can be parametrized by the map

$$\begin{aligned} j_f : \mathbb{R}^n &\longrightarrow \mathbb{R}^{2n} \\ (q_1, \cdots, q_n) &\longmapsto (q_1, \cdots, q_n, p_1 = f_1(q), \cdots, p_n = f_n(q)). \end{aligned}$$

Let us denote $\alpha = \sum_i p_i dq_i$ and so $\omega = d\alpha$. Then L_f is a Lagrangian submanifold if and only

$$j_f^*(\omega) = 0 \iff d(j_f^*(\alpha)) = 0 \iff j_f^*(\alpha) = dF \quad \text{for some} \quad F = F(q).$$

Here we have used Poincaré Lemma in the last step. Since $j_f^*\alpha = \sum_i f_i(q)dq_i$, we find that L_f is a Lagrangian submanifold if and only there exists a function $F(q)$ such that

$$f_i(q) = \frac{\partial F(q)}{\partial q_i}, \qquad i = 1, \cdots, n.$$

Again, such function F is called the ***generating function*** *of the Lagrangian submanifold L. In general, any Lagrangian submanifold will look locally like this example in Darboux coordinate.*

3.6 Moment Map

We will move on to discuss Hamiltonian systems with symmetries. This is formulated in terms of the concept of moment map, which is a geometric generalization of the linear and angular momentum. Such symmetry will help us to reduce the Hamiltonian system to lower dimensional phase spaces.

3.6.1 Lie Group and Lie Algebra

The symmetry we are concerned is continuous symmetry, which is described by Lie groups.

Definition 3.6.1. A **Lie group** is a smooth manifold which also has a group structure. That is a smooth manifold G with group operators $\cdot$ such that

$$\begin{aligned} G \times G &\longrightarrow G & G &\longrightarrow G \\ (g_1, g_2) &\longmapsto g_1 \cdot g_2 & g &\longmapsto g^{-1} \end{aligned}$$

are all smooth maps. The dimension $\dim G$ is the dimension of G as a manifold.

Example 3.6.2.

- $(\mathbb{R}, +)$ *is a Lie group with group operators*

$$\begin{aligned} \mathbb{R} \times \mathbb{R} &\longrightarrow \mathbb{R} & \mathbb{R} &\longrightarrow \mathbb{R} \\ (x, y) &\longmapsto x + y & x &\longmapsto -x \end{aligned}$$

- $(S^1 = \left\{e^{i\theta}\right\}, \cdot)$ *is a Lie group with group operators*

$$\begin{aligned} S^1 \times S^1 &\longrightarrow S^1 & S^1 &\longmapsto S^1 \\ (e^{i\theta_1}, e^{i\theta_2}) &\longmapsto e^{i(\theta_1+\theta_2)} & e^{i\theta} &\longmapsto e^{-i\theta} \end{aligned}$$

- $GL_n(\mathbb{R})$, *$n \times n$ invertible real matrices, with group operation by matrix multiplication.*

$$\dim(GL_n(\mathbb{R})) = n^2.$$

- $SL_n(\mathbb{R}) = \{A \in GL_n(\mathbb{R}) \mid \det(A) = 1\}$.

$$\dim SL_n(\mathbb{R}) = n^2 - 1.$$

- $O_n(\mathbb{R}) = \left\{A \in GL_n(\mathbb{R}) \mid A^T A = 1\right\} =$ *orthogonal linear transformations on* $\mathbb{R}^n$.
- $SO_n(\mathbb{R}) = \{A \in O_n(\mathbb{R}) \mid \det A = 1\} =$ *special orthogonal linear transformations on* $\mathbb{R}^n$.

$$\dim(O_n(\mathbb{R})) = \dim(SO_n(\mathbb{R})) = \frac{1}{2}n(n-1).$$

Definition 3.6.3. An action of a Lie group G on a manifold M is a homomorphism of manifolds

$$\begin{aligned} \rho : G \times M &\longrightarrow M \\ (g, m) &\longmapsto \rho_g(m) \xlongequal{\text{denoted by}} g \cdot m \end{aligned}$$

which it itself defines a group action. It is a smooth action if ρ is a smooth map.

Example 3.6.4. *The S^1-rotation on the plane $\mathbb{R}^2 \simeq \mathbb{C}$*

$$\begin{aligned} S^1 \times \mathbb{C} &\longrightarrow \mathbb{C} \\ (e^{i\theta}, z) &\longmapsto e^{i\theta} z \end{aligned}$$

is a smooth action. In (x, y)-coordinate, this is

$$\begin{pmatrix} x \\ y \end{pmatrix} \longmapsto \begin{pmatrix} \cos\theta & -\sin\theta \\ \sin\theta & \cos\theta \end{pmatrix} \begin{pmatrix} x \\ y \end{pmatrix}.$$

Example 3.6.5. *Let V be a vector field on M. Assume the flow ρ_t of V exists for all t, hence we obtain a map*

$$\begin{aligned} \mathbb{R} \times M &\longrightarrow M \\ (t, x) &\longmapsto \rho_t(x). \end{aligned}$$

This is a smooth $\mathbb{R}$-action, since we have

$$\rho_t \circ \rho_s = \rho_{t+s}.$$

Definition 3.6.6. For a G-action on M,

- the orbit of G through $x \in M$ is

$$G \cdot x := \{\rho_g(x) \,|\, g \in G\} \subset M.$$

- the stabilizer of $x \in M$ is the subgroup

$$G_x := \{g \in G \,|\, \rho_g(x) = x\} \subset G.$$

- the quotient space (called orbit space)

$$M/G = \text{the space of all orbits.}$$

- the G-action is transitive if there is only one orbit, i.e.,

$$M/G = \text{single point.}$$

- the G-action is free if for any $x \in M$,

$$G_x = 1.$$

Example 3.6.7. *There is a natural free S^1-action on S^3 as follows. Let us identify*

$$S^3 = \left\{(z_1, z_2) \in \mathbb{C}^2 \;\middle|\; |z_1|^2 + |z_2|^2 = 1\right\}.$$

Then the S^1-action is the natural rotation on the phases of z_1, z_2

$$e^{i\theta} : (z_1, z_2) \longmapsto (e^{i\theta} z_1, e^{i\theta} z_2).$$

The orbit space is $S^3/S^1 \simeq \mathbb{CP}^1 \simeq S^2$.

Let G be a Lie group. We can associate a linear space $\mathfrak{g}$, called its Lie algebra. Roughly speaking, the Lie group G represents continuous transformations, and the Lie algebra $\mathfrak{g}$ represents infinitesimal transformations. There are different ways to express $\mathfrak{g}$,

$$\mathfrak{g} = T_e G = \{\text{left-invariant vector fields on } G\}.$$

Here $e \in G$ is the identity. One essential structure that inherits the non-linear nature of G is the Lie bracket

$$[-,-]: \mathfrak{g} \times \mathfrak{g} \longrightarrow \mathfrak{g}$$

which satisfies the following basic properties

① bilinear.

② skew-symmetry: $[a,b] = -[b,a]$.

③ Jacobi identity: $[[a,b],c] + [[b,c],a] + [[c,a],b] = 0$.

Example 3.6.8. *Let $G = GL_n(\mathbb{R})$. Then*

$$\mathfrak{g} = gl_n(\mathbb{R}) = \{n \times n \text{ real matrices}\}.$$

For a matrix $A \in gl_n(\mathbb{R})$, the following path

$$\gamma(t) = I + tA \in GL_n(\mathbb{R}) \qquad \text{for } t \text{ small}$$

represents a curve on $GL_n(\mathbb{R})$ with $\gamma(0) = I$. The corresponding tangent vector is

$$\left.\frac{d}{dt}\right|_{t=0} \gamma(t) = A \in T_e G,$$

as described. The Lie bracket is the usual commutator

$$[A,B] = AB - BA, \qquad A, B \in gl_n(\mathbb{R}).$$

Example 3.6.9. *For a Lie group $G \subset GL_n(\mathbb{R})$, we will have*

$$\mathfrak{g} \subset gl_n(\mathbb{R}).$$

The Lie bracket will be induced from the commutator in $gl_n(\mathbb{R})$.

- $G = SL_n(\mathbb{R})$. $\mathfrak{g} = sl_n(\mathbb{R}) = \{A \in gl_n(\mathbb{R}) \mid \operatorname{Tr} A = 0\}$.

  We can see this by considering a curve

  $$\gamma(t) = 1 + tA + O(t^2) \in SL_n(\mathbb{R}).$$

  Then $\det(\gamma(t)) = 1 + t \operatorname{Tr} A + \cdots$, *which requires* $\operatorname{Tr} A = 0$.

- $G = SO_n(\mathbb{R})$. $\mathfrak{g} = so_n(\mathbb{R}) = \left\{A \in gl_n(\mathbb{R}) \mid A + A^T = 0\right\}$.

  Again, we can understand this by looking at the first order

  $$(1 + tA)(1 + tA^T) = 1 + t(A + A^T) + O(t^2).$$

Assume we have a G-action on a manifold M. For every element $A \in \mathfrak{g}$, we can get a vector field $V_A \in \mathrm{Vect}(M)$ representing the infinitesimal action. Let $\gamma(t)$ be a path on G with

$$\gamma(0) = e, \qquad \gamma'(0) = A.$$

Then at any point $x \in M$, the action $\gamma(t) \cdot x$ gives a path through x, and we define

$$V_A(x) = \left.\frac{d}{dt}\right|_{t=0} (\gamma(t) \cdot x).$$

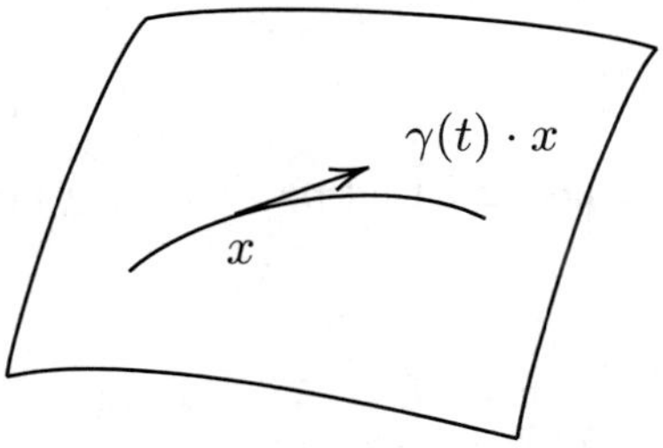

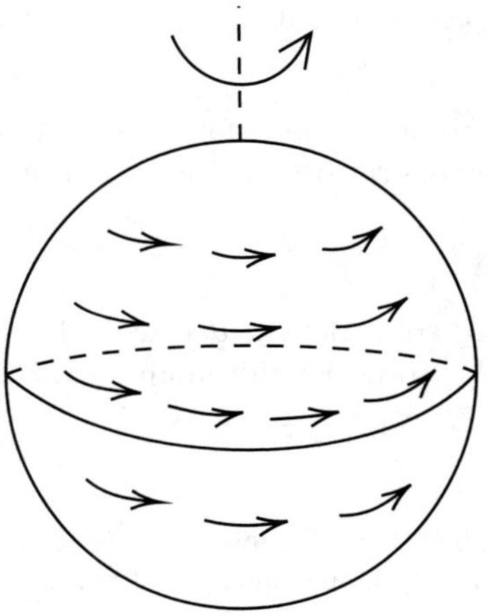

Therefore the G-action defines a map

$$\begin{aligned} \mathfrak{g} &\longrightarrow \mathrm{Vect}(M) \\ A &\longmapsto V_A \end{aligned}$$

Proposition 3.6.10. *The map $\mathfrak{g} \to \mathrm{Vect}(M)$ is a Lie algebra morphism, i.e., it preserves the bracket*

$$[V_A, V_B] = V_{[A,B]}, \qquad \forall A, B \in \mathfrak{g}.$$

Here $[V_A, V_B]$ is the Lie bracket for vector fields.

Remark. Let $\mathrm{Diff}(M)$ denote the group of diffeomorphisms from M to itself. This can be viewed as an infinite dimensional Lie group, whose Lie algebra (infinitesimal transformation) is $\mathrm{Vect}(M)$. A G-action on M gives a group homomorphism

$$G \longrightarrow \mathrm{Diff}(M).$$

Passing to tangent map at identity, we find the above map

$$\mathfrak{g} \longrightarrow \mathrm{Vect}(M).$$

3.6.2 Moment Map

Definition 3.6.11. Let (M, ω) be a symplectic manifold. A diffeomorphism

$$\varphi : M \longrightarrow M$$

is called a **symplectomorphism** if it preserves ω, i.e.,

$$\varphi^* \omega = \omega.$$

We now consider a Lie group G acting on a symplectic manifold (M, ω) by symplectomorphism:

$$g^* \omega = \omega, \qquad \forall g \in G.$$

At the infinitesimal level, let $A \in \mathfrak{g}$ and $V_A \in \mathrm{Vect}(M)$ be the vector field of infinitesimal transformation generated by A. Then

$$\mathscr{L}_{V_A} \omega = 0,$$

or equivalently, $d\iota_{V_A}\omega = 0$. Now we are looking for a better situation by taking all V_A to be Hamiltonian vector fields, so

$$\iota_{V_A}\omega = df_A \qquad \text{for } f_A \in C^\infty(M).$$

Note that f_A is not unique, and is determined up to a shift of constant. We consider a situation where we do can fix such ambiguity in a coherent way. This motivates the following definition.

Definition 3.6.12. The action G on a symplectic manifold (M, ω) via symplectomorphism is called **Hamiltonian** if there is a map

$$\mathscr{H} : \mathfrak{g} \longrightarrow C^\infty(M)$$

such that
① $\iota_{V_A}\omega = d\mathscr{H}_A$.
② $\mathscr{H}$ preserves the Poisson bracket:

$$\{\mathscr{H}_A, \mathscr{H}_B\} = \mathscr{H}_{[A,B]}.$$

Here ① can be also represented by a commutative diagram

$$\begin{array}{ccccccc} A & \in & \mathfrak{g} & \longrightarrow & C^\infty(M) & \ni & \mathscr{H}_A \\ & & & \searrow & \downarrow & & \downarrow \text{Hamiltonian vector field} \\ & & & & \mathrm{Vect}(M) & \ni & V_A \end{array}$$

Now we consider a Hamiltonian group action as above. The map

$$\mathscr{H} : \mathfrak{g} \longrightarrow C^\infty(M)$$

can be viewed as giving a linear set of functions on M. Dually, we can think about these functions as defining a map (called **moment map**)

$$\mu : M \longrightarrow \mathfrak{g}^*.$$

For any element $A \in \mathfrak{g}$,

$$\langle \mu(-), A \rangle = \mathscr{H}_A(-).$$

Here $\langle -, - \rangle$ is the natural pairing between $\mathfrak{g}^*$ and $\mathfrak{g}$.

Example 3.6.13. *Let* $G = \underbrace{S^1 \times \cdots \times S^1}_{n}$. *Its Lie algebra is* $\mathfrak{g} = \mathbb{R}^{\oplus n}$ *with trivial bracket (*G *is abelian). Then a moment map for* G *acting on* (M, ω) *is described by* n *functions* $\mathscr{H}_1, \cdots, \mathscr{H}_n \in C^\infty(M)$ *which are pairwise Poisson commuting:*

$$\{\mathscr{H}_i, \mathscr{H}_j\} = 0, \qquad \forall i, j.$$

In general, moment map encode constants of motion of the Hamiltonian system from symmetries.

Example 3.6.14. *Consider the translation action of* $\mathbb{R}^n$ *on the phase space* $\mathbb{R}^{2n}$

$$\vec{a} \cdot (\vec{x}, \vec{p}) = (\vec{x} + \vec{a}, \vec{p})$$

where $\vec{a} \in \mathbb{R}^n$, $(\vec{x}, \vec{p}) \in \mathbb{R}^{2n}$. *The Lie algebra is*

$$\mathfrak{g} = \mathbb{R}^n = \mathbb{R}e_1 \oplus \cdots \oplus \mathbb{R}e_n,$$

where e_i *represents the generator of translation along* x^i. *The corresponding vector field is*

$$V_{e_i} = \frac{\partial}{\partial x^i}.$$

For the symplectic form $\omega = \sum_i dp^i \wedge dx^i$, *we have*

$$\iota_{V_{e_i}} \omega = df_i,$$

where $f_i = -p_i$. *Then the map*

$$\begin{aligned} \mu : \mathbb{R}^{2n} &\longrightarrow \mathbb{R}^n \\ (\vec{x}, \vec{p}) &\longmapsto -\vec{p} = (-p_1, \cdots, -p_n) \end{aligned}$$

gives the corresponding moment map.

Example 3.6.15. *We consider rotation* S^1 *on* $\mathbb{R}^2$

$$e^{i\theta} : \begin{pmatrix} x \\ y \end{pmatrix} \longmapsto \begin{pmatrix} \cos\theta & -\sin\theta \\ \sin\theta & \cos\theta \end{pmatrix} \begin{pmatrix} x \\ y \end{pmatrix}$$

or in complex variable $z = x + iy$,

$$e^{i\theta} : z \longmapsto e^{i\theta} z.$$

The symplectic form is $\omega = dx \wedge dy$.

$$(e^{i\theta})^*(dx \wedge dy) = d(\cos\theta\ x - \sin\theta\ y) \wedge d(\sin\theta\ x + \cos\theta\ y) = dx \wedge dy$$

So S^1 *acts on* $\mathbb{R}^2$ *via symplectomorphism. The Lie algebra is*

$$\mathfrak{g} = \mathbb{R}e,$$

where the generator $e = \frac{\partial}{\partial\theta}$. *Let us compute the corresponding vector field* V_e *generated by* e.

$$V_e\Big|_{(x,y)} = \frac{\partial}{\partial\theta}\Big|_{\theta=0} \begin{pmatrix} \cos\theta & -\sin\theta \\ \sin\theta & \cos\theta \end{pmatrix} \begin{pmatrix} x \\ y \end{pmatrix} = \begin{pmatrix} 0 & -1 \\ 1 & 0 \end{pmatrix} \begin{pmatrix} x \\ y \end{pmatrix} = \begin{pmatrix} -y \\ x \end{pmatrix}.$$

So

$$V_e = -y\frac{\partial}{\partial x} + x\frac{\partial}{\partial y},$$

which is the vector field representing plane rotation. Now

$$\iota_{V_e}\omega = -ydy - xdx = d\left(-\frac{1}{2}x^2 - \frac{1}{2}y^2\right).$$

So the function

$$\mu(x, y) = -\frac{1}{2}\left(x^2 + y^2\right) + c = -\frac{1}{2}|z|^2 + c, \qquad c \in \mathbb{R} \text{ constant}$$

gives a moment map for the S^1*-rotation on* $\mathbb{R}^2$.

Example 3.6.16. *We consider the group* $GL_n(\mathbb{R})$ *acting on the phase space* $\mathbb{R}^{2n}$

$$O : \begin{pmatrix} \vec{x} \\ \vec{p} \end{pmatrix} \longmapsto \begin{pmatrix} O\vec{x} \\ (O^t)^{-1}\vec{p} \end{pmatrix}, \qquad O \in GL_n(\mathbb{R}).$$

Here $\vec{x} = \begin{pmatrix} x_1 \\ \vdots \\ x_n \end{pmatrix}$ *and* $\vec{p} = \begin{pmatrix} p_1 \\ \vdots \\ p_n \end{pmatrix}$. *It is clear that this action preserves the symplectic form*

$$\omega = \sum_{i=1}^{n} dp_i \wedge dx_i.$$

For every element $A \in \mathfrak{g} = gl_n(\mathbb{R})$, *it generates a vector field*

$$\begin{aligned} V_A &= \frac{d}{dt}\Big|_{t=0} \left(e^{tA}\vec{x}\right)^i \frac{\partial}{\partial x_i} + \frac{d}{dt}\Big|_{t=0} \left(e^{-tA^t}\vec{p}\right)^i \frac{\partial}{\partial p_i} \\ &= \sum_{ij} A_{ij}x_j\frac{\partial}{x_i} - \sum_{ij} A_{ij}p_i\frac{\partial}{\partial p_j}. \end{aligned}$$

This is a Hamiltonian vector field with Hamiltonian function $\mathscr{H}_A = -\sum_{ij} A_{ij} p_i x_j$. *This shows that the action is Hamiltonian with moment map*

$$\mu : \mathbb{R}^{2n} \longrightarrow \mathfrak{g}^*$$

where

$$\langle \mu(x,p), A \rangle = -\sum_{ij} A_{ij} p_i x_j.$$

If we consider the subgroup $SO_n(\mathbb{R})$ *which again gives a Hamiltonian action on* $\mathbb{R}^{2n}$. *Its Lie algebra* $so_n(\mathbb{R})$ *are skew-symmetric matrices, which are spanned by matrices of the form*

$$E_{ij} = \begin{pmatrix} 0 & & & & 0 \\ & \ddots & & 1 & \\ & & 0 & & \\ & -1 & & \ddots & \\ 0 & & & & 0 \end{pmatrix}$$

Here the only nonzero entries of the matrix E_{ij} *are* 1 *for the* (ij)*-component and* -1 *for the* (ji)*-component. The corresponding Hamiltonian is*

$$\mathscr{H}_{E_{ij}} = x_i p_j - x_j p_i.$$

When $n = 3$, *they are precisely components of the angular momentum.*

3.6.3 Symplectic Reduction

In a Hamiltonian system with a $2n$-dim phase space, if we find a continuous symmetry, then we can reduce it to a $(2n-2)$-dim problem. This works in great generality, for a Lie group symmetry, coined the name symplectic reduction.

Let us first describe the reduction process for a Hamiltonian system $(M, \omega, \mathscr{H})$, and f be a constant of motion. Its Hamiltonian vector field V_f generates a flow $\rho_t : M \to M$, which defines an $\mathbb{R}$-action

$$\begin{aligned} \mathbb{R} \times M &\longrightarrow M \\ (t,x) &\longmapsto \rho_t(x). \end{aligned}$$

The function f itself can be viewed as defining a moment map for the action

$$\mu = f : M \longrightarrow \mathbb{R}.$$

Suppose we choose local symplectic coordinates on $\mathcal{U} \subset M$

$$(q_1, \cdots, q_n, p_1, \cdots, p_n)$$

with $p_n = f$. Since f is a constant of motion,

$$\{\mathscr{H}, f\} = 0,$$

so

$$\frac{\partial \mathscr{H}}{\partial q_n} = 0 \qquad \Longrightarrow \qquad \mathscr{H} = \mathscr{H}(q_1, \cdots, q_{n-1}, p_1, \cdots, p_n).$$

Since $p_n = f$ is conserved, we set it to be a fixed value

$$p_n = c, \qquad c \in \mathbb{R}.$$

The motion is described by

$$\begin{cases} \dot{q}_i = \dfrac{\partial \mathscr{H}}{\partial p_i}(q_1, \cdots, q_{n-1}, p_1, \cdots, p_n, c), & i = 1, \cdots, n-1 \\ \dot{p}_i = -\dfrac{\partial \mathscr{H}}{\partial q_i}(q_1, \cdots, q_{n-1}, p_1, \cdots, p_n, c), & i = 1, \cdots, n-1 \\ \dot{q}_n = \dfrac{\partial \mathscr{H}}{\partial p_n} \\ \dot{p}_n = 0 \quad \text{and set } p_n = c. \end{cases}$$

The reduced phase space is

$$\mathscr{U}_{\text{red}} = \big\{(q_1, \cdots, q_{n-1}, p_1, \cdots, p_{n-1}) \in \mathbb{R}^{2n-2} \mid \\ (q_1, \cdots, q_{n-1}, a, p_1, \cdots, p_{n-1}, c) \in \mathscr{U} \text{ for some } a\big\}.$$

Different a's will be connected by the flow of

$$V_f = V_{p_n} = -\frac{\partial}{\partial q_n}.$$

Therefore we see that $\mathscr{U}_{\text{red}}$ is some orbit space. The reduced Hamiltonian is

$$\begin{aligned} \mathscr{H}_{\text{red}} : \mathscr{U}_{\text{red}} &\longrightarrow \mathbb{R} \\ (q_1, \cdots, q_{n-1}, p_1, \cdots, p_{n-1}) &\longmapsto \mathscr{H}(q_1, \cdots, q_{n-1}, p_1, \cdots, p_{n-1}, c) \end{aligned}$$

Now we end up with a phase space of dim $= 2n-2$, and the trajectories in $q_n(t)$, $p_n(t)$ are

$$\begin{cases} q_n(t) = \displaystyle\int^t \frac{\partial \mathscr{H}}{\partial p_n}\, dt \\ p_n(t) = c \end{cases}$$

If g is another constant of motion independent of f, then we can use g to reduce the phase space to $(2n-4)$-dimensional. In particular, in the Liouville integrable case where we find n independent constants of motion (see Section 4.1.1), we are reduced to a 0-dim phase space, and this will allow us to find trajectories of $q_i(t)$'s, $p_i(t)$'s.

Now we formulate the above process systematically. Let (M, ω) be a symplectic manifold. Let G be a Lie group acting on M via symplectomorphisms. Assume we have a moment map for the G-action

$$\mu : M \longrightarrow \mathfrak{g}^*$$

where $\mathfrak{g}$ is the Lie algebra of G. There is the adjoint G-action on $\mathfrak{g}$

$$\begin{aligned} Ad : G \times \mathfrak{g} &\longrightarrow \mathfrak{g} \\ (a, \xi) &\longmapsto Ad_a(\xi). \end{aligned}$$

For matrix Lie groups and Lie algebras, it is

$$Ad_a(\xi) = a\xi a^{-1}.$$

It induces a G-action on $\mathfrak{g}^*$ via duality

$$Ad_a(\varphi)(\xi) := \varphi(Ad_{a^{-1}}(\xi)), \qquad \forall \varphi \in \mathfrak{g}^*, \ \xi \in \mathfrak{g}.$$

The moment map is G-equivariant:

$$\mu(a \cdot m) = Ad_a \mu(m), \qquad \forall a \in G, \ m \in M.$$

In particular, if we consider the origin

$$0 \in \mathfrak{g}^*$$

which is itself a G-orbit, then we have an induced action

$$G \times \mu^{-1}(0) \longrightarrow \mu^{-1}(0).$$

Theorem 3.6.17 (Marsden-Weinstein-Meyer). *Suppose G is a compact Lie group, which defines a Hamiltonian action on the symplectic manifold (M, ω) with moment map $\mu : M \to \mathfrak{g}^*$. Assume G acts freely on $\mu^{-1}(0)$. Then*

(1) $M_{red} := \mu^{-1}(0)/G$ is a smooth manifold.

(2) $\pi : \mu^{-1}(0) \to M_{red}$ is a principal G-bundle.

(3) The 2-form $\omega|_{\mu^{-1}(0)}$ descends to define a symplectic form on M_{red}. Equivalently, there is a symplectic form ω_{red} on M_{red} such that

$$\pi^* \omega_{red} = \omega|_{\mu^{-1}(0)}.$$

We will call $(M_{\text{red}}, \omega_{\text{red}})$ the **symplectic quotient** of (M, ω). Sometimes it is denoted by

$$M_{\text{red}} = M//G.$$

The Marsden-Weinstein-Meyer Theorem says that we can use symmetries of the system to reduce it to something easier.

Example 3.6.18. *Consider the S^1-action on $\mathbb{C}^n$ by*

$$e^{i\theta} \cdot (z_1, \cdots, z_n) = (e^{i\theta} z_1, \cdots, e^{i\theta} z_n).$$

This is a Hamiltonian action with moment map

$$\mu(z) = -\frac{1}{2}(|z_1|^2 + \cdots + |z_n|^2) + c.$$

Here $c \in \mathbb{R}$ is an arbitrary constant. We have also used that the Lie algebra of S^1 is $\mathbb{R}$.

Let us set $c = \frac{1}{2}$ and consider

$$\mu(z) = -\frac{1}{2}(|z_1|^2 + \cdots + |z_n|^2) + \frac{1}{2}.$$

Now we have

$$\mu^{-1}(0) = S^{2n-1}$$

which is the unit sphere in $\mathbb{C}^n$. The symplectic quotient is

$$\mathbb{C}^n // S^1 = \mu^{-1}(0)/S^1 = S^{2n-1}/S^1 \simeq \mathbb{CP}^{n-1}.$$

The symplectic form ω_{red} is (up to a constant) the Fubini-Study symplectic form on $\mathbb{CP}^{n-1}$.

The Marsden-Weinstein-Meyer Theorem tells us how to reduce at the 0 level set of the moment map. We can also perform reduction at other level sets of μ. We will sketch the basic idea here. Let $\xi \in \mathfrak{g}^*$ and consider a coadjoint orbit

$$\mathcal{O} = G \cdot \xi \subset \mathfrak{g}^*.$$

Since $\mu : M \to \mathfrak{g}^*$ is G-equivariant, $G : \mu^{-1}(\mathcal{O}) \to \mu^{-1}(\mathcal{O})$ preserves the level set of the orbit $\mathcal{O}$. Assume G is compact and acts freely on $\mu^{-1}(\mathcal{O})$. Then it can be shown that the quotient

$$\mu^{-1}(\mathcal{O})/G$$

admits a symplectic manifold structure. In fact, this can be reduced to the case in Marsden-Weinstein-Meyer Theorem as follows. The coadjoint orbit $\mathcal{O}$ is equipped with the Kostant-Kirillov symplectic form $\omega_{\mathcal{O}}$. We consider the natural product action

$$G \curvearrowright M \times \mathcal{O} \qquad \text{with symplectic form } \omega_M - \omega_{\mathcal{O}}$$

which is Hamiltonian with moment map

$$\begin{aligned} \mu_{\mathcal{O}} : M \times \mathcal{O} &\longrightarrow \mathfrak{g}^* \\ (p, s) &\longmapsto \mu(p) - \xi \end{aligned}$$

Then $\mu^{-1}(\mathcal{O})/G = \mu_{\mathcal{O}}^{-1}(0)/G$.

Chapter 4

Integrable System

4.1 Liouville Integrability

4.1.1 Liouville Integrability and Liouville Tori

There is a special class of systems where solutions of the Hamilton's equations can be found by quadratures, *i.e.*, by solving a finite number of algebraic equations and definite integrals. This happens when there are enough many conserved quantities, and such dynamic systems are generally known as Liouville integrable systems.

Before we discuss the general Liouville integrable system, let us consider a simple example of the n-dim harmonic oscillator. The Hamiltonian is

$$\mathscr{H} = \sum_{i=1}^{n} \left(\frac{1}{2} p_i^2 + \frac{\omega_i^2}{2} q_i^2 \right) = \sum_{i=1}^{n} f_i,$$

where $f_i(q,p) = \frac{1}{2} p_i^2 + \frac{\omega_i^2}{2} q_i^2$ for $i = 1, \cdots, n$. The symplectic form is

$$\omega = \sum_i dp_i \wedge dq_i,$$

which leads to the Poisson bracket

$$\{q_i, p_i\} = \delta_{ij}.$$

The functions $f_1, \cdots, f_n$ are pairwise Poisson commuting

$$\{f_i, f_j\} = 0, \qquad \forall i, j.$$

In particular, f_i's are all constants of motion

$$\{f_i, \mathscr{H}\} = 0, \qquad \forall i.$$

Define the common level set

$$P_c = \left\{(q_i, p_i) \in \mathbb{R}^{2n} \mid f_i(q, p) = c_i,\ i = 1, \cdots, n\right\},$$

where $c_i > 0$ are constants. Then

$$P_c \simeq T^n = \underbrace{S^1 \times \cdots \times S^1}_{n}$$

is diffeomorphic to a n-dim torus T^n. These tori foliate the phase space and can be parametrized by n angles θ_i. They evolve linearly in time with frequencies ω_i under Hamilton's equations.

In fact, consider the change of variables

$$p_i = r_i \cos\theta_i, \qquad q_i = \frac{r_i}{\omega_i}\sin\theta_i.$$

Then $dp_i \wedge dq_i = \frac{1}{\omega_i} r_i dr_i \wedge d\theta_i = d\left(\frac{r_i^2}{2\omega_i}\right) \wedge d\theta_i$. Let $I_i = \frac{r_i^2}{2\omega_i}$. Then

$$(q_i, p_i) \longmapsto (\theta_i, I_i)$$

is a canonical transformation, and $\{\theta_i, I_j\} = \delta_{ij}$.

The Hamiltonian is $\mathscr{H} = \sum_i \omega_i I_i$ which only depend on I_i's. Hamilton's equations become

$$\begin{cases} \dot{I}_i = \{I_i, \mathscr{H}\} = 0 \\ \dot{\theta}_i = \{\theta_i, \mathscr{H}\} = \omega_i \end{cases}$$

The solution is $I_i =$ const and $\theta_i = \omega_i t +$const. We see the phase flow becomes a flow linearly in time on each tori T^n with frequencies ω_i.

Definition 4.1.1. A Hamiltonian system is a triple $(M, \omega, \mathscr{H})$ where (M, ω) is a symplectic manifold, and $\mathscr{H}$ is a smooth function on M called the Hamiltonian function. A function f on M is called a **constant of motion** (or **integral of motion**) if

$$\{f, \mathscr{H}\} = 0.$$

A set of functions $f_1, \cdots, f_k$ are said to be **independent** if their differentials $(df_1)_p, \cdots, (df_k)_p$ are linearly independent at all points p in some open dense subset of M.

If $f_1, f_2, \cdots, f_k$ are independent and pairwise Poisson commuting constants of motion, let V_{f_i} denote the corresponding Hamiltonian vector field, then

$$0 = \{f_i, f_j\} = \iota_{V_{f_i}} \iota_{V_{f_j}} \omega, \qquad \forall i, j$$

says that $\{V_{f_1}, \cdots, V_{f_k}\}$ span an isotropic subspace of T_pM at almost each p. This implies

$$k \leq \frac{1}{2}\dim M.$$

Definition 4.1.2. A Hamiltonian system $(M, \omega, \mathscr{H})$ is **(completely) integrable** in the **Liouville sense** if it processes $n = \frac{1}{2}\dim M$ independent constants of motion, $f_1, \cdots, f_n$, which are pairwise in involution with respect to the Poisson bracket, i.e., $\{f_i, f_j\} = 0$, $\forall i, j$.

Note that $\mathscr{H}$ is also a constant of motion, then we can express $\mathscr{H} = \mathscr{H}(f_1, \cdots, f_n)$ as a function of f_i's. Let us denote

$$f = (f_1, \cdots, f_n) : M \longrightarrow \mathbb{R}^n.$$

A point $c \in \mathbb{R}^n$ is called a regular value of f if $df_1, \cdots, df_n$ are linearly independent at all points of the level set $f^{-1}(c)$. In this case, $f^{-1}(c)$ is also a smooth manifold, and it turns out that any compact component of $f^{-1}(c)$ must be a torus. Those are called **Liouville tori**.

Lemma 4.1.3. *Let $f = (f_1, \cdots, f_n) : M \longrightarrow \mathbb{R}^n$ be as above from a completely integrable system. Assume c be a regular value of f. Then the level set $f^{-1}(c)$ is a Lagrangian submanifold (see Definition 3.5.20) of M.*

Proof. Let $L = f^{-1}(c)$ and we have $\dim L = n$. By construction, a tangent vector $V \in T_pM$ for $p \in L$ lies in T_pL if and only if $V(f_i) = 0$ at p for all $i = 1, \cdots, n$. Let $V_{f_i} \in \mathrm{Vect}(M)$ be the Hamiltonian vector field of the function f_i. Since f_i's are in involution, we have

$$V_{f_i}(f_j) = \{f_i, f_j\} = 0, \qquad \forall i, j.$$

So V_{f_i}'s are all tangent to L. For any $p \in L$, since df_i's are linearly independent at p, the vector fields V_{f_i} are also linearly independent at p. It follows that

$$T_pL = \mathrm{Span}\{V_{f_1}|_p, \cdots, V_{f_n}|_p\} \qquad \text{for each } p \in L.$$

Since $\omega(V_{f_i}, V_{f_j}) = \{f_i, f_j\} = 0$, we conclude that T_pL is a linear Lagrangian subspace of T_pM. Therefore $L = f^{-1}(c)$ is a Lagrangian submanifold of M. □

4.1.2 Liouville-Arnol'd Theorem

Theorem 4.1.4 (Liouville-Arnol'd Theorem)**.** *Let $(M, \omega, \mathscr{H})$ be a Liouville integrable system of $\dim = 2n$ with constants of motion $f_1, \cdots, f_n$ in involution. Let $c \in \mathbb{R}^{2n}$ be a regular value of $f = (f_1, \cdots, f_n)$. Then*

(1) *$f^{-1}(c)$ is a smooth manifold invariant under the Hamilton flow with $\mathscr{H} = \mathscr{H}(f_1, \cdots, f_n)$.*

(2) *Assume $f^{-1}(c)$ is compact and connected, then it is diffeomorphic to the n-dim torus*

$$T^n = \{(\varphi_1, \cdots, \varphi_n) \ \mathit{mod}\ 2\pi\}.$$

(3) *The motion on $f^{-1}(c)$ under the Hamiltonian flow is linear:*

$$\frac{d\varphi_i}{dt} = \omega_i(c), \qquad \omega_i(c) \textit{ are frequencies depending on } c.$$

(4) *There are coordinates* $I_1, \cdots, I_n$ *which are constants of motion such that* $\omega = \sum_i dI_i \wedge d\varphi_i$ *in a neighborhood of* $f^{-1}(c)$.

(5) *The equations of motion can be integrated by quadratures.*

Here $\omega = \sum_i dI_i \wedge d\varphi_i$ *says that* $\{\varphi_1, \cdots, \varphi_n, I_1, \cdots, I_n\}$ *form Darboux coordinates.* $\{\varphi_1, \cdots, \varphi_n\}$ *are called* ***angle coordinates*** *and* $\{I_1, \cdots, I_n\}$ *are called* ***action coordinates***.

For simplicity, we explain the action-angle variables for a Liouville integrable system on the standard phase space $(\mathbb{R}^{2n}, \omega = \sum_{i=1}^n dp_i \wedge dq_i)$. Let

$$f = (f_1, \cdots, f_n) : \mathbb{R}^{2n} \longrightarrow \mathbb{R}^n.$$

The motion occurs on an n-dimensional torus T^n as a level set of f:

$$\begin{cases} f_1(q,p) = c_1 \\ \quad\vdots \\ f_n(q,p) = c_n \end{cases}$$

We solve it for $p_i = p_i(c, q)$.

Let us choose a basis $\{\gamma_1, \cdots, \gamma_n\}$ of one-dimensional cycles on the torus $f^{-1}(c)$ depending continuously on c (locally).

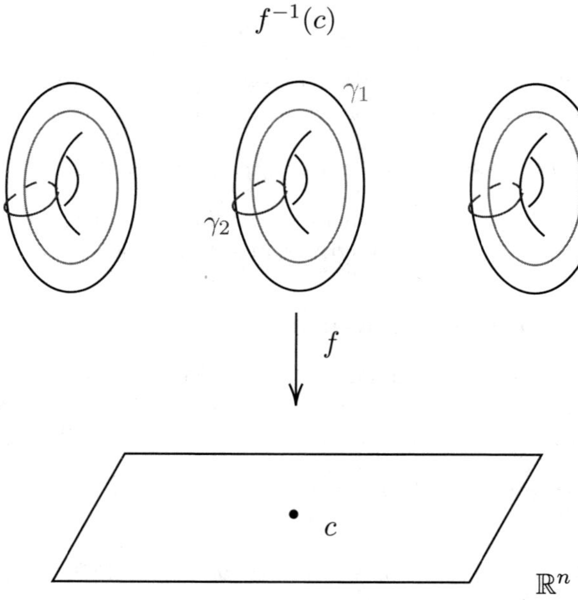

The action variables are given by

$$I_i(c) = \frac{1}{2\pi} \oint_{\gamma_i} \sum_k p_k(c, q) dq_k.$$

Since c's are constants of motion, so is $I_i(c)$.

Assume I_i's are independent functions of c_i's, and so the map $c_i \mapsto I_i(c)$ can be inverted. Now we can construct a canonical transformation

$$(q_i, p_i) \longmapsto (\varphi_i, I_i)$$

as follows. Consider the generating function

$$S(I, q) = \int_{q_0}^{q} \sum_k p_k(\tilde{q}, I) d\tilde{q}.$$

Here we have expressed p_k as a function of q, I. The integration $\int_{q_0}^{q}$ is a path lies inside $f^{-1}(c)$ from an arbitrary fixed point q_0 to q.

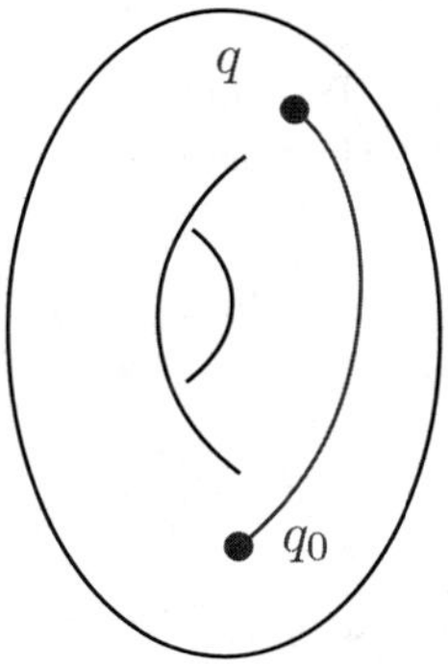

Fact. This integral $\int_{q_0}^{q}$ is invariant under a continuous deformation of the path, and will jump by $\oint_{\gamma_i} pdq = 2\pi I_i$ if we choose another path that will wind around the circle γ_i. So $S(I, q)$ is in fact a multi-valued function with periods $2\pi I_i$. This follows from Stokes Theorem (Example 3.1.28) and the fact that the Liouville tori is a Lagrangian submanifold (Lemma 4.1.3). Thus the symplectic 2-form $\sum_i dp_i \wedge dq_i$ being zero when restricted to a Lagrangian submanifold implies that the 1-form $\sum_i p_i dq_i$ is closed when restricted to the Liouville tori (see also Example 3.5.21).

Introduce

$$\varphi_i = \frac{\partial S}{\partial I_i}.$$

Then φ_i is multi-valued with period 2π, hence can be viewed as a coordinate on S^1.

Consider the total differential

$$dS = \sum_i \frac{\partial S}{\partial q_i} dq_i + \sum_i \frac{\partial S}{\partial I_i} dI_i = \sum_i p_i dq_i + \sum_i \varphi_i dI_i.$$

Applying d again, we find

$$0 = d^2 S = \sum_i dp_i \wedge dq_i + \sum_i d\varphi_i \wedge dI_i,$$

i.e.,

$$\sum_i dp_i \wedge dq_i = \sum_i dI_i \wedge d\varphi_i.$$

Therefore $(q_i, p_i) \mapsto (\varphi_i, I_i)$ is a canonical transformation, and S is the corresponding generating function. In the action-angle variable (φ_i, I_i), the Hamiltonian $\mathscr{H}$ can be expressed as a function of c_i's, hence a function of I_i's.

$$\mathscr{H} = \mathscr{H}(I).$$

Therefore the Hamilton's equation becomes

$$\begin{cases} \dot{I}_i = \{I_i, \mathscr{H}\} = 0 \\ \dot{\varphi}_i = \{\varphi_i, \mathscr{H}\} = \dfrac{\partial \mathscr{H}}{\partial I_i} = \omega_i \end{cases}$$

So $I_i = \text{const}$ and $\varphi_i = \omega_i t + \text{const}$. In other words, the motion is given by a flow on tori with linear dependence on the time t. The Hamilton's equations are now solved via quadratures.

Example 4.1.5 (1-dim Harmonic Oscillator). *$\mathscr{H} = \frac{1}{2}p^2 + \frac{\omega^2}{2}q^2$ on $\mathbb{R}^2$. This is integrable with $f_1 = \mathscr{H}$. The torus is*

$$\gamma_c : \frac{1}{2}p^2 + \frac{\omega^2}{2}q^2 = c.$$

We can solve $p = p(q, c) = \pm\sqrt{2c - \omega^2 q^2}$. The action variable is

$$I = \frac{1}{2\pi}\oint_{\gamma_c} \sqrt{2c - \omega^2 q^2}\, dq = \frac{c}{\omega}.$$

The generating function reads

$$S(I, q) = \int^q \sqrt{2I\omega - \omega^2 x^2}\, dx.$$

Therefore the angle variable is

$$\theta(I, q) = \frac{\partial S}{\partial I} = \omega \int^q \frac{dx}{\sqrt{2I\omega - \omega^2 x^2}} = \arcsin\left(\sqrt{\frac{\omega}{2I}}q\right).$$

Inverting this, we find

$$q = \sqrt{\frac{2I}{\omega}} \sin\theta.$$

This gives the full canonical transformation

$$\begin{cases} q = \sqrt{\dfrac{2I}{\omega}} \sin\theta \\ p = \sqrt{2I\omega - \omega^2 q^2} = \sqrt{2I\omega}\cos\theta \end{cases}$$

This is the same as we find in the beginning of this subsection.

4.2 Integrability of Kepler Problem

4.2.1 Complete Integrability

We revisit the Kepler problem of motions in $\mathbb{R}^3$ with central conservative forces. Let

$$(q_1, q_2, q_3) \in \mathbb{R}^3$$

denote the position, and (p_1, p_2, p_3) be the conjugate momentum. The phase space is $\mathbb{R}^6$ with symplectic form

$$\omega = dp_1 \wedge dq_1 + dp_2 \wedge dq_2 + dp_3 \wedge dq_3$$

and Poisson relations $\{q_i, p_j\} = \delta_{ij}$. The Hamiltonian is given by (we set $m = 1$ for convenience):

$$\mathscr{H} = \frac{1}{2} \sum_{i=1}^{3} p_i^2 + V(r),$$

where $r = \sqrt{q_1^2 + q_2^2 + q_3^2}$. Since $V(r)$ is rotationally invariant, Noether theorem implies the conservation of the angular momentum $\vec{J} = \vec{q} \times \vec{p}$. The components

$$J_i = \sum_{j,k} \epsilon_{ijk} q_j p_k$$

satisfies the following Poisson bracket relations

$$\{J_i, J_j\} = \epsilon_{ijk} J_k.$$

Consider $J^2 = J_1^2 + J_2^2 + J_3^2$. We can compute

$$\{J^2, J_1\} = \{J^2, J_2\} = \{J^2, J_3\} = 0.$$

Therefore we have three independent commuting constants of motion in involution, namely

$$\mathscr{H}, \qquad J_3, \qquad J^2.$$

This puts the Kepler problem in the class of Liouville integrable systems.

4.2.2 Action-Angle Variables

Let us apply Liouville-Arnol'd in this case. Firstly, it would be convenient to change to the spherical coordinates

$$\begin{cases} q_1 = r \sin\theta \cos\phi \\ q_2 = r \sin\theta \sin\phi \\ q_3 = r \cos\theta \end{cases}$$

To find the canonical transformation, we simply require

$$\alpha = \sum_i p_i dq_i = p_r dr + p_\theta d\theta + p_\phi d\phi.$$

This allows us to find the relations

$$\begin{cases} p_1 = \dfrac{1}{r}\left(rp_r\cos\phi\sin\theta + p_\theta\cos\theta\cos\phi - p_\phi\dfrac{\sin\phi}{\sin\theta}\right) \\ p_2 = \dfrac{1}{r}\left(rp_r\sin\phi\sin\theta + p_\theta\cos\theta\sin\phi + p_\phi\dfrac{\cos\phi}{\sin\theta}\right) \\ p_3 = p_r\cos\theta - \dfrac{1}{r}p_\theta\sin\theta \end{cases}$$

In sphere coordinates, we find

$$\begin{cases} \mathscr{H} = \dfrac{1}{2}\left(p_r^2 + \dfrac{1}{r^2}p_\theta^2 + \dfrac{1}{r^2\sin^2\theta}p_\phi^2\right) + V(r) \\ J^2 = p_\theta^2 + \dfrac{1}{\sin^2\theta}p_\phi^2 \\ J_3 = p_\phi \end{cases}$$

Now we use E to denote the total energy, so

$$\mathscr{H} = E.$$

We still use J^2, J_3 for the value of the corresponding constants of motion. We can solve

$$\begin{cases} p_r = \sqrt{2(E - V(r)) - \dfrac{J^2}{r^2}} \\ p_\theta = \sqrt{J^2 - \dfrac{J_3^2}{\sin^2\theta}} \\ p_\phi = J_3 \end{cases}$$

Note that the variables (r, θ, ϕ) and their conjugate momentums (p_r, p_θ, p_ϕ) are completely separated. The trajectories in the (r, p_r), (θ, p_θ) and (ϕ, p_ϕ) phase planes are pictured as below.

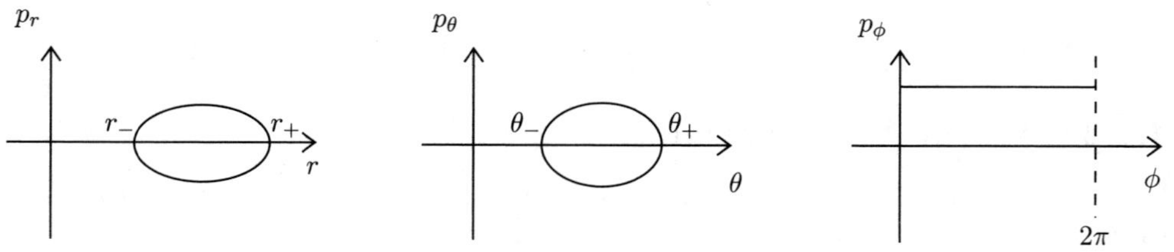

Here $r_\pm$ are roots of the equation

$$V(r) + \frac{J^2}{2r^2} = E.$$

Similarly $\theta_\pm$ are given by

$$\sin\theta_\pm = \pm J_3/J.$$

The ϕ-degree is simply the circle parametrized by ϕ.

The action variables are given by

$$\begin{cases} I_r = \dfrac{1}{2\pi} \oint \sqrt{2(E - V(r)) - \dfrac{J^2}{r^2}} dr \\ I_\theta = \dfrac{1}{2\pi} \oint \sqrt{J^2 - \dfrac{J_3^2}{\sin^2\theta}} d\theta \\ I_\phi = \dfrac{1}{2\pi} \oint J_3 d\phi \end{cases}$$

Here the integrations are over the corresponding circle trajectory in the phase space.

- The integral for I_ϕ is trivial. We find

$$\boxed{I_\phi = J_3}$$

Without lost of generality, we assume

$$J_3 > 0.$$

- The integral for I_θ can be performed. Let

$$\cos\alpha = J_3/J.$$

Then

$$I_\theta = \frac{J}{2\pi} \oint \sqrt{1 - \frac{\cos^2\alpha}{\sin^2\theta}} d\theta.$$

The integration is over the following circle

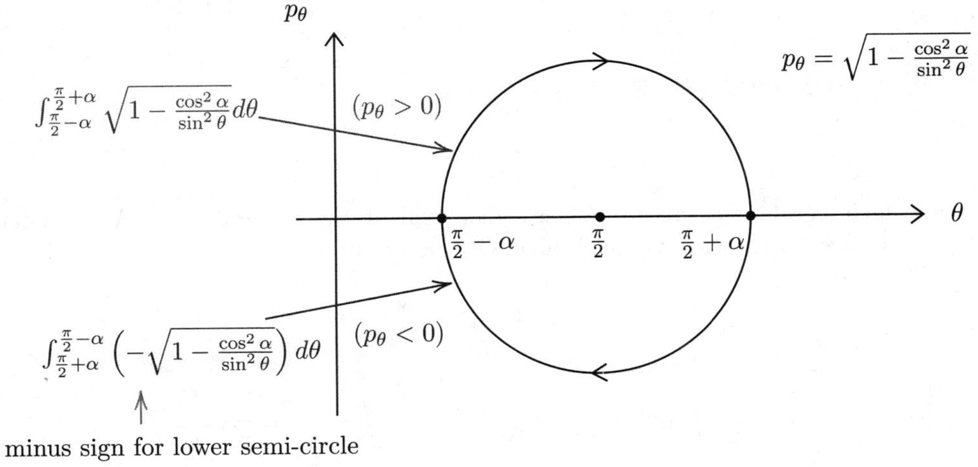

$$I_\theta = \frac{J}{2\pi}\oint \frac{d\theta}{\sqrt{1-\frac{\cos^2\alpha}{\sin^2\theta}}} - \frac{J_3}{2\pi}\oint \frac{\cos\alpha d\theta}{\sin^2\theta\sqrt{1-\frac{\cos^2\alpha}{\sin^2\theta}}}.$$

For the first integral, let

$$\sin x = -\frac{\cos\theta}{\sin\alpha}, \qquad x = \arcsin\left(-\frac{\cos\theta}{\sin\alpha}\right).$$

$$\frac{dx}{d\theta} = \frac{1}{\sqrt{1-\frac{\cos^2\alpha}{\sin^2\theta}}}.$$

So

$$\begin{aligned}\oint \frac{d\theta}{\sqrt{1-\frac{\cos^2\alpha}{\sin^2\theta}}} &= \int_{\frac{\pi}{2}-\alpha}^{\frac{\pi}{2}+\alpha} \frac{d\theta}{\sqrt{1-\frac{\cos^2\alpha}{\sin^2\theta}}} + \int_{\frac{\pi}{2}+\alpha}^{\frac{\pi}{2}-\alpha} \frac{d\theta}{-\sqrt{1-\frac{\cos^2\alpha}{\sin^2\theta}}} \\ &= 2\int_{\frac{\pi}{2}-\alpha}^{\frac{\pi}{2}+\alpha} \frac{d\theta}{\sqrt{1-\frac{\cos^2\alpha}{\sin^2\theta}}} \\ &= 2\int_{-\frac{\pi}{2}}^{\frac{\pi}{2}} dx = 2\pi.\end{aligned}$$

Similarly, let

$$\sin y = -\cot\alpha\,\cot\theta, \qquad y = \arcsin(-\cot\alpha\,\cot\theta).$$

$$\frac{dy}{d\theta} = \frac{\cos\alpha}{\sin^2\theta}\frac{1}{\sqrt{1-\frac{\cos^2\alpha}{\sin^2\theta}}}.$$

Then

$$\oint \frac{\cos\alpha d\theta}{\sin^2\theta\sqrt{1-\frac{\cos^2\alpha}{\sin^2\theta}}} = 2\int_{\frac{\pi}{2}-\alpha}^{\frac{\pi}{2}+\alpha} \frac{\cos\alpha d\theta}{\sin^2\theta\sqrt{1-\frac{\cos^2\alpha}{\sin^2\theta}}} = 2\int_{-\frac{\pi}{2}}^{\frac{\pi}{2}} dy = 2\pi.$$

Put everything together, we find

$$I_\theta = J - J_3$$

- The integral for I_r can not be performed explicitly for general $V(r)$. We consider the case for the gravitational force with potential

$$V(r) = -\frac{k}{r}.$$

$$I_r = \frac{1}{2\pi}\oint \sqrt{2E + \frac{2k}{r} - \frac{J^2}{r^2}}dr = \frac{1}{2\pi}\oint \frac{2E + \frac{2k}{r} - \frac{J^2}{r^2}}{\sqrt{2E + \frac{2k}{r} - \frac{J^2}{r^2}}}dr$$

$$= \frac{1}{2\pi}\oint \frac{2E+\frac{k}{r}}{\sqrt{2E+\frac{2k}{r}-\frac{J^2}{r^2}}}dr + \frac{1}{2\pi}\oint \frac{\frac{k}{r}}{\sqrt{2E+\frac{2k}{r}-\frac{J^2}{r^2}}}dr$$
$$- \frac{1}{2\pi}\oint \frac{\frac{J^2}{r^2}}{\sqrt{2E+\frac{2k}{r}-\frac{J^2}{r^2}}}dr$$
$$= \frac{1}{2\pi}\oint d(\sqrt{2Er^2+2kr-J^2}) + \frac{1}{2\pi}\oint \frac{\frac{k}{r}}{\sqrt{2E+\frac{2k}{r}-\frac{J^2}{r^2}}}dr$$
$$- \frac{1}{2\pi}\oint \frac{\frac{J^2}{r^2}}{\sqrt{2E+\frac{2k}{r}-\frac{J^2}{r^2}}}dr.$$

The first integral is zero, and the last two integrals can be performed explicitly. We present the answer here. Assume $E < 0$ so we are in the situation of bounded region. Let

$$a = \frac{k}{-2E}, \qquad e = \sqrt{1+2J^2E/k^2}.$$

Consider the change of coordinate

$$r/a = 1 - e\cos x \qquad \text{and} \qquad r/a = \frac{1-e^2}{1+e\cos y},$$

we find

$$I_r = \frac{1}{2\pi}\sqrt{\frac{k^2}{-2E}}\oint dx - J\oint dy.$$

By the similar discussion as before, we get

$$I_r = \sqrt{\frac{k^2}{-2E}} - J$$

As a result, we can express the energy E in terms of the action variables as

$$\mathscr{H} = E = -\frac{k^2}{2(I_r+J)^2} = -\frac{k^2}{2(I_r+I_\theta+I_\phi)^2}.$$

In summary, we have

$$\begin{cases} E = -\dfrac{k^2}{2(I_r+I_\theta+I_\phi)^2} \\ J = I_\theta + I_\phi \\ J_3 = I_\phi \end{cases}$$

- The generating function is

$$S = \int^r \sqrt{2(E-V(r)) - \frac{J^2}{r^2}}dr + \int^\theta \sqrt{J^2 - \frac{J_3^2}{\sin^2\theta}}d\theta + \int^\phi J_3 d\phi.$$

The corresponding angle variables are

$$\varphi_r = \frac{\partial S}{\partial I_r}, \qquad \varphi_\theta = \frac{\partial S}{\partial I_\theta}, \qquad \varphi_\phi = \frac{\partial S}{\partial I_\phi}.$$

Remark. The above result is for unit mass $m = 1$. In general, we will find

$$\begin{cases} I_r = \sqrt{\dfrac{mk^2}{-2E}} - J \\ I_\theta = J - J_3 \\ I_\phi = J_3 \end{cases}$$

and the inverse map is

$$\begin{cases} \mathscr{H} = E = -\dfrac{mk^2}{2(I_r + I_\theta + I_\phi)^2} \\ J = I_\theta + I_\phi \\ J_3 = I_\phi \end{cases}$$

The Hamilton's equations are

$$\begin{cases} \dot{\varphi}_r = \omega_r = \dfrac{\partial \mathscr{H}}{\partial I_r} \\ \dot{\varphi}_\theta = \omega_\theta = \dfrac{\partial \mathscr{H}}{\partial I_\theta} \\ \dot{\varphi}_\phi = \omega_\phi = \dfrac{\partial \mathscr{H}}{\partial I_\phi} \end{cases}$$

We find that in the case for $V(r) = -\frac{k}{r}$, the three frequencies ω_r, ω_θ, ω_ϕ are equal:

$$\omega_r = \omega_\theta = \omega_\phi = \frac{mk^2}{(I_r + I_\theta + I_\phi)^3}.$$

Note that $E = -\frac{mk^2}{2(I_r+I_\theta+I_\phi)^2}$, so

$$\omega_r = \omega_\theta = \omega_\phi = \frac{m}{|k|}\left(\frac{-2E}{m}\right)^{\frac{3}{2}}.$$

Let τ denote the period of the orbit. Then

$$\frac{m}{k}\left(\frac{-2E}{m}\right)^{\frac{3}{2}} = \frac{2\pi}{\tau}.$$

Using $a = -\frac{k}{2E}$ which is the semi-major axis of the elliptical orbit, we find

$$\left(\frac{\tau}{2\pi}\right)^2 = \frac{m}{k}a^3.$$

This is Kepler's third law.

4.3 Hamilton-Jacobi v.s. Liouville Integrability

4.3.1 Local Complete Solution

We discuss the connection between the Hamilton-Jacobi equation and Liouville integrability. We will see that these two methods of solving the dynamics are essentially equivalent.

Let us consider a time-independent Hamiltonian system $\mathscr{H}(p,q)$ on an open subset of the standard phase space $\mathbb{R}^{2d}$. Recall that the Hamilton-Jacobi equation is

$$\frac{\partial \widehat{W}}{\partial t} + \mathscr{H}\left(q, \frac{\partial \widehat{W}}{\partial q}\right) = 0.$$

Here $\widehat{W} = W(q,t)$ is a function of q's and t. In the time-independent case, the Hamiltonian is conserved and let $\mathscr{H} = E$ be the energy. We can set

$$\widehat{W}(q,t) = W(q) - Et.$$

So the Hamilton-Jacobi equation becomes

$$\mathscr{H}\left(q, \frac{\partial W}{\partial q}\right) = E \qquad (*)$$

and we treat E as one of the constants of motion. We will focus on this case and mainly consider the Hamilton-Jacobi equation in the above form $(*)$.

Definition 4.3.1. A local complete solution of Hamilton-Jacobi equation is a function $W = W(\tilde{p}, q)$ and $E = E(\tilde{p})$, where $\tilde{p}$, q are in some open subset of $\mathbb{R}^d$, such that

① $\mathscr{H}\left(q, \frac{\partial W}{\partial q}(\tilde{p},q)\right) = E(\tilde{p})$.

② $\det\left(\frac{\partial^2 W}{\partial \tilde{p}_i \partial q_i}\right) \neq 0$ on the domain of definition for $(\tilde{p}, q)$.

Condition ① says that we obtain a family of solutions of Hamilton-Jacobi equation parameterized by $\tilde{p}$. Let us consider condition ②. Let $p_i = \frac{\partial W}{\partial q_i}(\tilde{p}, q)$. Then ② says that locally the transformation $(q,\tilde{p}) \mapsto \left(q, p = \frac{\partial W}{\partial q_i}(\tilde{p},q)\right)$ can be inverted (Implicit function Theorem)

$$(q,p) \longmapsto (q,\tilde{p}).$$

Therefore a local complete solution $W(\tilde{p},q)$ gives a family that essentially depends on d parameters. We can define similarly

$$\tilde{q}_i = \frac{\partial W}{\partial \tilde{p}_i}(\tilde{p}, q).$$

Then the same reasoning shows that

$$(q,\tilde{p}) \longmapsto (\tilde{q},\tilde{p})$$

can also be inverted locally. As a result, the map

$$(q,p) \longmapsto (\tilde{q},\tilde{p})$$

defines a local change of coordinates on the phase space.

Proposition 4.3.2. *Let $(W(\tilde{p},q), E(\tilde{p}))$ be a local complete solution of the Hamilton-Jacobi equation. Let*

$$\tilde{q}_i = \frac{\partial W}{\partial \tilde{p}_i}(\tilde{p},q), \qquad p_i = \frac{\partial W}{\partial q_i}(\tilde{p},q).$$

Then the transformation $(q,\tilde{p}) \mapsto (\tilde{q},\tilde{p})$ defines a (local) canonical transformation with W being the generating function.

Proof. For $W = W(\tilde{p},q)$, we have

$$dW = \frac{\partial W}{\partial \tilde{p}} d\tilde{p} + \frac{\partial W}{\partial q} dq = \sum_i \tilde{q}_i d\tilde{p}_i + \sum_i p_i dq_i.$$

Applying d again, we find

$$\sum_i dp_i \wedge dq_i = \sum_i d\tilde{p}_i \wedge d\tilde{q}_i.$$ □

In the $(\tilde{q},\tilde{p})$-coordinate, the Hamiltonian is

$$\mathscr{H} = E = E(\tilde{p}).$$

Hamilton's equations become

$$\dot{\tilde{p}}_i = 0, \qquad \dot{\tilde{q}}_i = \frac{\partial E}{\partial \tilde{p}_i},$$

and all motions are linear. The upshot is that finding a local complete solution of the Hamilton-Jacobi equation reduces the problem of solving the equations of motion via quadratures.

4.3.2 Integrability via Hamilton-Jacobi Theory

Proposition 4.3.3. *The Hamilton-Jacobi equation has a local complete solution near some point in the 2d-dimensional phase space if and only if the Hamiltonian system possesses d independent constants of motion in involution near that point.*

Proof. ($\Rightarrow$) By the proposition above, we can find local canonical transformation

$$(q,p) \longmapsto (\tilde{q},\tilde{p})$$

such that the Hamiltonian $\mathscr{H} = \mathscr{H}(\tilde{p})$. It is clear that $\tilde{p}_1 \cdots, \tilde{p}_d$ gives d independent constants of motion in involution: $\{\tilde{p}_i, \tilde{p}_j\} = 0$.

($\Leftarrow$) Let $f = (f_1, \cdots, f_d)$ be d independent constants of motion in involution locally near a point z_*. The Hamiltonian is a function of f_i's

$$\mathscr{H} = E(f_1, \cdots, f_d).$$

So E is the energy function we are looking for. Next we try to find a system of local canonical coordinates and to solve the Hamilton-Jacobi equation.

Let us first choose a local canonical coordinate system (q, p)

$$\omega = \sum_i dp_i \wedge dq_i$$

such that p_i's are transversal to the level set of f around the point z_*. This is always possible locally. This also means that (q_i, f_i) form a local coordinate system around z_*.

Let $\mathcal{U}$ be a small neighborhood of $f(z_*)$ in $\mathbb{R}^d$. Let

$$\hat{q} : \mathcal{U} \longrightarrow \mathbb{R}^d$$

be some chosen map valued in domain as described below. Let $c_1, \cdots, c_d$ be coordinates on $\mathcal{U}$.

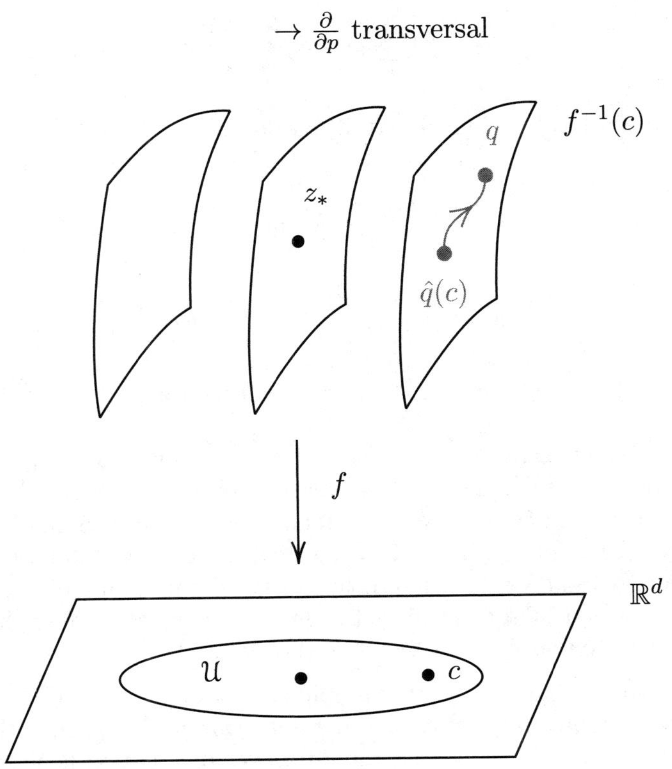

Define the following function of c_i and q_i

$$W(c,q) = \int_{\hat{q}(c)}^{q} \sum_i p_i dq_i \Big|_{f^{-1}(c)}.$$

Here the integral is over any path on the level set $f^{-1}(c)$ connecting the chosen initial point $\hat{q}(c)$ and the endpoint q. Note that since (q_i, f_i) is a local coordinate, points on the level set $f^{-1}(c)$ is locally uniquely determined by the value of q.

Using the fact that the 1-form $\sum_i p_i dq_i$ is a closed 1-form when restricting to $f^{-1}(c)$, we see that $W(c,q)$ is locally well-defined and does not depend on the choice of the path locally. We can also use the coordinate transformation

$$(q_i, p_i) \longmapsto (q_i, c_i = f_i)$$

to express $p_i = p_i(c,q)$. This allows us to write

$$W(c,q) = \int_{\hat{q}(c)}^{q} \sum_i p_i(c,q) dq_i.$$

In particular, we have $\frac{\partial W}{\partial q_i}(c,q) = p_i(c,q)$. So

$$dW = \sum_i p_i dq_i + \sum_i \frac{\partial W}{\partial c_i} dc_i$$

and W generates a local canonical transformation

$$(q_i, p_i) \longmapsto \left(\frac{\partial W}{\partial c_i}, c_i \right).$$

The Hamiltonian is

$$\mathscr{H} = \mathscr{H}\left(q, \frac{\partial W}{\partial q}(c,q) \right) = E(c).$$

So $(W(c,q), E(c))$ gives a local complete solution of the Hamilton-Jacobi equation. □

Remark. Liouville integrability says geometrically the level set $f^{-1}(c)$ form a family of Lagrangian submanifolds in the phase space parametrized by c. We know that in local canonical coordinates (q,p), a Lagrangian submanifold can be always described by $p_i = \frac{\partial W}{\partial q_i}$ (see Example 3.5.21). Then the local complete solution of the Hamilton-Jacobi equation gives an explicit description of the d-dim family of Lagrangians $p_i = \frac{\partial W}{\partial q_i}(c,q)$ parametrized by c. This gives the geometric explanation of the relationship between the Hamilton-Jacobi and Liouville theory.

In Liouville integrable system, we can find action-angle variables (I_i, φ_i). The I_i's are functions of c_i's but in general not the same. However, we can still use c_i's for a canonical transformation to solve the Hamilton-Jacobi equation hence the system.

Example 4.3.4. *Consider the Kepler problem*

$$\begin{cases} \mathscr{H} = E = \frac{1}{2}\left(p_r^2 + \frac{1}{r^2}p_\theta^2 + \frac{1}{r^2\sin^2\theta}p_\phi^2\right) + V(r) \\ J^2 = p_\theta^2 + \frac{1}{\sin^2\theta}p_\phi^2 \\ J_3 = p_\phi \end{cases}$$

We consider a simplified situation that the motion lies on the xy-plane. This is to set

$$\theta = \frac{\pi}{2}, \qquad p_\theta = 0.$$

This does not lose much since conservation of angular momentum forces the motion to lie on a plane, and we can use rotation symmetry to set the plane to be xy. We can also check that setting $\theta = \frac{\pi}{2}$, $p_\theta = 0$ is consistent with the Hamilton's equations. In fact,

$$\begin{cases} \dot\theta = \{\theta, \mathscr{H}\} = \frac{1}{r^2}p_\theta \\ \dot p_\theta = \frac{\cos\theta}{r^2\sin^3\theta}p_\phi^2 \end{cases}$$

$\theta = \frac{\pi}{2}$, $p_\theta = 0$ obviously satisfies these equations. So we now consider the 2-dim problem with

$$\begin{cases} \mathscr{H} = E = \frac{1}{2}\left(p_r^2 + \frac{1}{r^2}J^2\right) + V(r) \\ J = J_3 = p_\phi \end{cases}$$

with E and J being two independent constants of motion in involution. We have

$$\begin{cases} p_r = \sqrt{2(E - V(r)) - \frac{J^2}{r^2}} \\ p_\phi = J \end{cases}$$

It is clear that we have change of coordinates

$$(r, \phi, p_r, p_\phi) \longleftrightarrow (r, \phi, E, J)$$

and therefore p_r, p_ϕ is transversed to level sets of E, J. A local complete solution of the Hamilton-Jacobi equation is

$$W(E, J, r, \phi) = \int^r \sqrt{2(E - V(u)) - \frac{J^2}{u^2}}\, du + \int^\phi J.$$

It leads to a canonical transformation

$$(r, \phi, p_r, p_\phi) \longmapsto (\alpha_E, \alpha_J, E, J)$$

where

$$\begin{cases} \alpha_E = \dfrac{\partial W}{\partial E} = \displaystyle\int^r \frac{du}{\sqrt{2(E - V(u)) - \frac{J^2}{u^2}}} \\ \alpha_J = \dfrac{\partial W}{\partial J} = \phi - \displaystyle\int^r \frac{J du}{u^2\sqrt{2(E - V(u)) - \frac{J^2}{u^2}}} \end{cases}$$

Since $\mathscr{H} = E$, the Hamilton's equations are

$$\dot{\alpha}_E = 1, \qquad \dot{\alpha}_J = 0.$$

Then

$$\begin{cases} t - t_0 = \displaystyle\int^r \frac{du}{\sqrt{2(E - V(u)) - \frac{J^2}{u^2}}} \\ \phi(r) = \displaystyle\int^r \frac{J du}{u^2\sqrt{2(E - V(u)) - \frac{J^2}{u^2}}} \end{cases}$$

Assume the case of gravitational potential

$$V(r) = -\frac{k}{r}.$$

The above integral can be performed and

$$\phi(r) = \arccos\left(\frac{J/r - k/J}{\sqrt{2E + k^2/J^2}}\right).$$

Let $l = J^2/k$, $e = \sqrt{1 + \frac{2EJ^2}{k^2}}$. We find the equation of orbit

$$\boxed{r = \frac{l}{1 + e\cos\phi}}$$

This gives the solution of Kepler problem as we have obtained before.

4.4 Toda Lattice

4.4.1 Toda Equations in Lax Form

The Toda lattice is a simple model for one dimensional crystal in solid state physics. It is also one of the earliest examples of non-linear completely integrable system. It exhibits soliton solutions to a chain of particles with non-linear interactions between nearest neighbors.

Consider n particles of unit mass arranged along a line at positions $q_1, \cdots, q_n$. We consider a force between each pair of adjacent particles, whose magnitude depends exponentially on the distance between them. Let p_i's denote the corresponding conjugate momentums. The Hamiltonian of the so-called Toda lattice is

$$\mathscr{H} = \frac{1}{2}\sum_{i=1}^{n} p_i^2 + \sum_{i=1}^{n-1} e^{-(q_{i+1}-q_i)}.$$

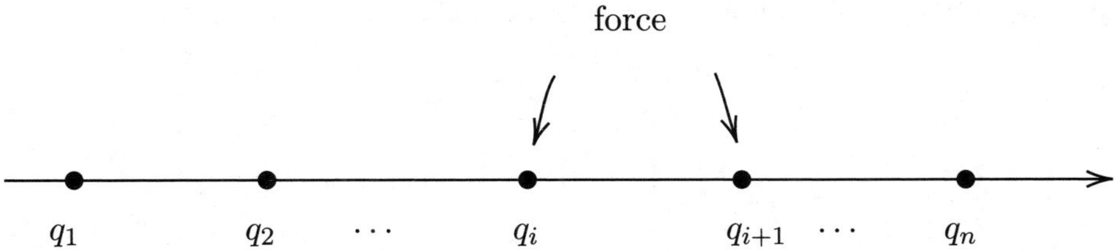

The Hamilton's equations read

$$\begin{cases} \dot{q}_i = \{q_i, \mathscr{H}\} = p_i \\ \dot{p}_i = \{p_i, \mathscr{H}\} = \begin{cases} -e^{q_1-q_2} & i=1 \\ e^{q_{i-1}-q_i} - q^{q_i-q_{i+1}} & 2 \le i \le n-1 \\ e^{q_{n-1}-q_n} & i=n \end{cases} \end{cases}$$

It turns out that the Hamilton's equation of Toda lattice can be put into Lax form, which will lead to Liouville integrability. Introduce a new set of variables

$$\begin{cases} a_i = \frac{1}{2} e^{\frac{q_i-q_{i+1}}{2}} & i = 1, \cdots, n-1 \\ b_i = -\frac{1}{2} p_i & i = 1, \cdots, n \end{cases}$$

The overall translation of q_i's is a symmetry and $\mathscr{H}$ only depends on a_i, b_i's. The evolution becomes

$$\begin{cases} \dot{a}_i = a_i(b_{i+1} - b_i) & i = 1, \cdots, n-1 \\ \dot{b}_i = 2(a_i^2 - a_{i-1}^2) & i = 1, \cdots, n \end{cases}$$

where we use the convention that $a_0 = a_n = 0$.

In these variables, the Hamiltonian has the form

$$\mathscr{H}(a, b) = 2\sum_{i=1}^{n} b_i^2 + 4\sum_{i=1}^{n-1} a_i^2.$$

We can compute the Poisson bracket between a_i, b_i's, and find

$$\{a_i, b_j\} = -\frac{1}{4}\delta_{ij} a_i + \frac{1}{4}\delta_{i,j-1} a_i, \qquad i = 1, \cdots, n-1, \ j = 1, \cdots, n,$$

while all the other entries are equal to zero. This Poisson bracket is degenerate in the a, b variables (i.e., there exists a non-constant function of a_i, b_i's which is Poisson-commuting with everything). In fact, the total momentum

$$\{b_1 + \cdots + b_n, a_i\} = \{b_1 + \cdots + b_n, b_i\} = 0.$$

For example,

$$\{a_2, b_2\} = -\frac{1}{4}a_2, \qquad \{a_2, b_3\} = \frac{1}{4}a_2.$$

These two terms will cancel out. Introduce the following $n \times n$ matrices

$$L = \begin{pmatrix} b_1 & a_1 & & & 0 \\ a_1 & b_2 & a_2 & & \\ & a_2 & \ddots & \ddots & \\ & & \ddots & \ddots & a_{n-1} \\ 0 & & & a_{n-1} & b_n \end{pmatrix}, \qquad A = \begin{pmatrix} 0 & a_1 & & & 0 \\ -a_1 & 0 & a_2 & & \\ & -a_2 & \ddots & \ddots & \\ & & \ddots & \ddots & a_{n-1} \\ 0 & & & -a_{n-1} & 0 \end{pmatrix}.$$

Proposition 4.4.1. *The Hamilton's equation for Toda lattice*

$$\begin{cases} \dot{a}_i = a_i(b_{i+1} - b_i) & i = 1, \cdots, n-1 \\ \dot{b}_j = 2(a_j^2 - a_{j-1}^2) & j = 1, \cdots, n \end{cases}$$

is equivalent to the following equation

$$\frac{dL}{dt} = [A, L].$$

Here the $n \times n$ matrices L and A are defined above.

Proof. This is a direct computation. Let us do a sample computation in the case $n = 3$.

$$L = \begin{pmatrix} b_1 & a_1 & 0 \\ a_1 & b_2 & a_2 \\ 0 & a_2 & b_3 \end{pmatrix}, \qquad A = \begin{pmatrix} 0 & a_1 & 0 \\ -a_1 & 0 & a_2 \\ 0 & -a_2 & 0 \end{pmatrix}.$$

$$\begin{aligned}
[A, L] &= AL - LA \\
&= \begin{pmatrix} 0 & a_1 & 0 \\ -a_1 & 0 & a_2 \\ 0 & -a_2 & 0 \end{pmatrix} \begin{pmatrix} b_1 & a_1 & 0 \\ a_1 & b_2 & a_2 \\ 0 & a_2 & b_3 \end{pmatrix} - \begin{pmatrix} b_1 & a_1 & 0 \\ a_1 & b_2 & a_2 \\ 0 & a_2 & b_3 \end{pmatrix} \begin{pmatrix} 0 & a_1 & 0 \\ -a_1 & 0 & a_2 \\ 0 & -a_2 & 0 \end{pmatrix} \\
&= \begin{pmatrix} a_1^2 & a_1 b_2 & a_1 a_2 \\ -a_1 b_1 & -a_1^2 + a_2^2 & a_2 b_3 \\ -a_1 a_2 & -a_2 b_2 & -a_2^2 \end{pmatrix} - \begin{pmatrix} -a_1^2 & a_1 b_1 & a_1 a_2 \\ -a_1 b_2 & a_1^2 - a_2^2 & a_2 b_2 \\ -a_1 a_2 & -a_2 b_3 & a_2^2 \end{pmatrix} \\
&= \begin{pmatrix} 2a_1^2 & a_1(b_2 - b_1) & 0 \\ a_1(b_2 - b_1) & 2a_2^2 - 2a_1^2 & a_2(b_3 - b_2) \\ 0 & a_2(b_3 - b_2) & -2a_2^2 \end{pmatrix}
\end{aligned}$$

as required. Therefore the Toda system equations can be put into the standard Lax form. □

4.4.2 Integrability of Toda Lattice

As we discussed in Section 1.5.5, for evolution equations in Lax form, the quantities

$$f_k = \frac{4}{k+1}\operatorname{Tr} L^{k+1}, \qquad k = 0, 1, \cdots, n-1$$

are constants of motion. Note that

$$f_0 = 4\operatorname{Tr} L = 4(b_1 + b_2 + \cdots + b_n)$$

is the total momentum (up to a constant).

$$\begin{aligned} f_1 = 2\operatorname{Tr} L^2 &= 2\operatorname{Tr} \begin{pmatrix} b_1 & a_1 & & & 0 \\ a_1 & b_2 & a_2 & & \\ & a_2 & \ddots & \ddots & \\ & & \ddots & \ddots & a_{n-1} \\ 0 & & & a_{n-1} & b_n \end{pmatrix} \begin{pmatrix} b_1 & a_1 & & & 0 \\ a_1 & b_2 & a_2 & & \\ & a_2 & \ddots & \ddots & \\ & & \ddots & \ddots & a_{n-1} \\ 0 & & & a_{n-1} & b_n \end{pmatrix} \\ &= 2\sum_i b_i^2 + 4\sum_i a_i^2 \end{aligned}$$

is precisely the Hamiltonian.

Observe that we have asymptotic behavior

$$f_k = \frac{4}{k+1}\left(\sum_{i=1}^{n} b_i^{k+1}\right) + O(a_i)$$

at regions where a_i's are small. The leading terms $\left\{\sum_{i=1}^{n} b_i^{k+1}\right\}_{k=0,\cdots,n-1}$ are independent functions at regions when b_i's are different. This follows from a standard Vandermonde determinant computation. Therefore $f_0, f_1, \cdots, f_k$ are independent in some open region of the phase space. We next show that these constants of motion are in involution, leading to Liouville integrability.

Lemma 4.4.2. *The eigenvalues of L are all distinct.*

We will leave this as an exercise.

We now show that the eigenvalues $\lambda_1 < \lambda_2 < \cdots < \lambda_n$ of L are in involution. This will prove that f_k's are in involution since

$$f_k = \frac{4}{k+1}\sum_{i=1}^{n} \lambda_i^{k+1}.$$

Let λ, μ be two eigenvalues of L, with normalized eigenvectors $\vec{v} = (v_1, \cdots, v_n)$ and $\vec{w} = (w_1, \cdots, w_n)$. Then

$$\begin{aligned} L\vec{v} &= \lambda\vec{v}, \qquad \langle v, v\rangle = 1, \\ L\vec{w} &= \mu\vec{w}, \qquad \langle w, w\rangle = 1. \end{aligned}$$

Here $\langle -, - \rangle$ is the Euclidean inner product. Recall

$$a_i = \frac{1}{2} e^{\frac{q_i - q_{i+1}}{2}}, \qquad b_i = -\frac{1}{2} p_i.$$

Let us compute

$$\begin{aligned} \frac{\partial \lambda}{\partial p_i} &= \frac{\partial}{\partial p_i} \langle \vec{v}, L\vec{v} \rangle = \left\langle \frac{\partial \vec{v}}{\partial p_i}, L\vec{v} \right\rangle + \left\langle \vec{v}, L \frac{\partial \vec{v}}{\partial p_i} \right\rangle + \left\langle \vec{v}, \frac{\partial L}{\partial p_i} \vec{v} \right\rangle \\ &= \left\langle \frac{\partial \vec{v}}{\partial p_i}, L\vec{v} \right\rangle + \left\langle L\vec{v}, \frac{\partial \vec{v}}{\partial p_i} \right\rangle + \left\langle \vec{v}, \frac{\partial L}{\partial p_i} \vec{v} \right\rangle \\ &= \lambda \left\langle \frac{\partial \vec{v}}{\partial p_i}, \vec{v} \right\rangle + \lambda \left\langle \vec{v}, \frac{\partial \vec{v}}{\partial p_i} \right\rangle + \left\langle \vec{v}, \frac{\partial L}{\partial p_i} \vec{v} \right\rangle = \left\langle \vec{v}, \frac{\partial L}{\partial p_i} \vec{v} \right\rangle. \end{aligned}$$

Here we have used $\frac{\partial}{\partial p_i} \langle \vec{v}, \vec{v} \rangle = \frac{\partial}{\partial p_i}(1) = 0$. Similarly, we have

$$\frac{\partial \lambda}{\partial q_i} = \left\langle \vec{v}, \frac{\partial L}{\partial q_i} \vec{v} \right\rangle.$$

Now

$$\frac{\partial L}{\partial p_i} = \begin{pmatrix} 0 & & & & & & \\ & \ddots & & & & 0 & \\ & & 0 & & & & \\ & & & -\frac{1}{2} & & & \\ & & & & 0 & & \\ & 0 & & & & \ddots & \\ & & & & & & 0 \end{pmatrix} \leftarrow i,$$

$$\uparrow$$
$$i$$

$$\frac{\partial L}{\partial q_i} = \begin{pmatrix} 0 & & & & & & \\ & \ddots & & & & 0 & \\ & & 0 & -\frac{1}{2} a_{i-1} & & & \\ & & -\frac{1}{2} a_{i-1} & 0 & \frac{1}{2} a_i & & \\ & & & \frac{1}{2} a_i & 0 & & \\ & 0 & & & & \ddots & \\ & & & & & & 0 \end{pmatrix}.$$

It follows that

$$\begin{cases} \dfrac{\partial \lambda}{\partial p_i} = \left\langle \vec{v}, \dfrac{\partial L}{\partial p_i} \vec{v} \right\rangle = -\dfrac{1}{2} v_i^2 \\ \dfrac{\partial \lambda}{\partial q_i} = \left\langle \vec{v}, \dfrac{\partial L}{\partial q_i} \vec{v} \right\rangle = a_i v_i v_{i+1} - a_{i-1} v_i v_{i-1} \end{cases}$$

Here we have set $a_0 = a_n = 0$. The same relations hold for the eigenvalue μ. Then one has

$$\begin{aligned}\{\lambda,\mu\} &= \sum_{i=1}^{n}\left(\frac{\partial\lambda}{\partial q_i}\frac{\partial\mu}{\partial p_i} - \frac{\partial\lambda}{\partial p_i}\frac{\partial\mu}{\partial q_i}\right)\\ &= \sum_{i=1}^{n}\left[(a_i v_i v_{i+1} - a_{i-1}v_i v_{i-1})\left(-\frac{1}{2}w_i^2\right) - \left(-\frac{1}{2}v_i^2\right)(a_i w_i w_{i+1} - a_{i-1}w_i w_{i-1})\right]\\ &= \frac{1}{2}\sum_{i=1}^{n} v_i w_i\left[a_i(v_i w_{i+1} - v_{i+1}w_i) + a_{i-1}(v_{i-1}w_i - v_i w_{i-1})\right]\end{aligned}$$

Let $R_i = a_i(v_i w_{i+1} - v_{i+1}w_i)$, $i = 0, \cdots, n$ with $R_0 = R_n = 0$. Then

$$\{\lambda,\mu\} = \frac{1}{2}\sum_{i=1}^{n} v_i w_i(R_i + R_{i-1}).$$

Consider the matrix

$$E_i = \begin{pmatrix} 0 & & & & & \\ & \ddots & & & 0 & \\ & & 0 & & & \\ & & & 1 & & \\ & & & & 0 & \\ & 0 & & & & \ddots & \\ & & & & & & 0 \end{pmatrix} \leftarrow i$$

$$\uparrow_{i}$$

We have

$$[E_i, L] = \begin{pmatrix} 0 & & & & & & \\ & \ddots & & & & 0 & \\ & & 0 & -a_{i-1} & & & \\ & & a_{i-1} & 0 & a_i & & \\ & & & -a_i & 0 & & \\ & 0 & & & & \ddots & \\ & & & & & & 0 \end{pmatrix} \leftarrow i$$

$$\uparrow_{i}$$

So

$$\langle\vec{v}, [E_i, L]\vec{w}\rangle = a_i(v_i w_{i+1} - v_{i+1}w_i) - a_{i-1}(v_{i-1}w_i - v_i w_{i-1}) = R_i - R_{i-1}.$$

On the other hand, using $L\vec{w} = \mu\vec{w}$, $L\vec{v} = \lambda\vec{v}$,

$$\langle\vec{v}, [E_i, L]\vec{w}\rangle = \langle\vec{v}, E_i L\vec{w}\rangle - \langle\vec{v}, LE_i\vec{w}\rangle = \langle\vec{v}, E_i L\vec{w}\rangle - \langle L\vec{v}, E_i\vec{w}\rangle = (\mu - \lambda)v_i w_i.$$

Thus

$$v_i w_i = \frac{R_i - R_{i-1}}{\mu - \lambda}.$$

It follows that

$$\{\lambda, \mu\} = \frac{1}{2}\sum_{i=1}^{n} v_i w_i (R_i + R_{i-1}) = \frac{1}{2(\mu - \lambda)}\sum_{i=1}^{n}(R_i^2 - R_{i-1}^2) = \frac{R_n^2 - R_0^2}{2(\mu - \lambda)} = 0.$$

We have now established the Liouville integrability of the Toda lattice.

4.5 Calogero-Moser System

4.5.1 Calogero-Moser Space via Symplectic Reduction

Calogero-Moser system is a one-dimensional many-body problem that can be explicitly solved. The Hamiltonian of Calogero-Moser system is

$$\mathscr{H} = \sum_{i=1}^{n} p_i^2 + \sum_{i \neq j} U(x_i - x_j).$$

Here the potential U has several forms.

- Rational Calogero-Moser system:

  $$U(r) = \frac{1}{r^2}.$$

  This system corresponds to particles on a line with an inverse square of the distance potential between any pair.

- Trigonometric Calogero-Moser system, also Sutherland system:

  $$U(r) = \frac{1}{4\sin^2(r/2)}.$$

  This system can be viewed as particles on the circle.

Let $M_n(\mathbb{C}) = \{n \times n\ \mathbb{C} - \text{valued matrices}\}$. We consider the phase space M with positions parametrized by $M_n(\mathbb{C})$ and their conjugate momentum.

$$M = T^* M_n(\mathbb{C}) = \underset{X=(X_{ij})}{\underset{\Psi}{M_n(\mathbb{C})}} \oplus \underset{Y=(Y_{ij})}{\underset{\Psi}{M_n(\mathbb{C})}}$$

Precisely speaking, we are considering system over $\mathbb{C}$. The (complex) symplectic form is

$$\omega = \operatorname{Tr} dX \wedge dY.$$

Consider the Lie group of the projective general linear group

$$G = PGL_n(\mathbb{C}) = GL_n(\mathbb{C})/\sim$$

where we identify an invertible complex matrix A with

$$A \sim \lambda A \qquad \text{for } \lambda \in \mathbb{C}^*.$$

Its Lie algebra can be identified with

$$sl_n(\mathbb{C}) = \{A \in M_n(\mathbb{C}) \mid \operatorname{Tr} A = 0\}.$$

In particular, $\dim_{\mathbb{C}} G = n^2 - 1$.

We consider the G-action on M by

$$g \cdot (X, Y) = (gXg^{-1}, gYg^{-1}).$$

It is clear that $g^*\omega = \omega$, so defines a symplectomorphism. This G-action has a moment map

$$\mu : M \longrightarrow sl_n(\mathbb{C})^*$$

as follows. First, we can define an inner product

$$\begin{aligned} \langle -, - \rangle : sl_n(\mathbb{C}) \times sl_n(\mathbb{C}) &\longrightarrow \mathbb{C} \\ (A, B) &\longmapsto \operatorname{Tr} AB \end{aligned}$$

This is a non-degenerate pairing and defines an isomorphism

$$\begin{aligned} sl_n(\mathbb{C}) &\longrightarrow sl_n(\mathbb{C})^* \\ A &\longmapsto \langle A, - \rangle = \operatorname{Tr}(A-) \end{aligned}$$

This isomorphism is compatible with the $PGL_n(\mathbb{C})$-action, since

$$gAg^{-1} \longmapsto \langle gAg^{-1}, - \rangle = \operatorname{Tr}(gAg^{-1}(-)) = \operatorname{Tr}(Ag^{-1}(-)g) = \langle A, g^{-1}(-)g \rangle.$$

Now we describe the moment map under the above identification

$$\begin{aligned} \mu : M &\longrightarrow sl_n(\mathbb{C})^* \simeq sl_n(\mathbb{C}) \\ (X, Y) &\longmapsto [X, Y] = XY - YX \end{aligned}$$

Exercise: Show that this is indeed the moment map for the action

$$PGL_n(\mathbb{C}) \curvearrowright M = T^* M_n(\mathbb{C}).$$

The coadjoint orbit related to the Calogero-Moser model is the orbit $\mathcal{O}$ through the point

$$\gamma = \operatorname{diag}(-1, -1, \cdots, -1, n-1) \in sl_n(\mathbb{C}).$$

Let us first describe explicitly elements of $\mathcal{O}$.

Proposition 4.5.1. *The orbit*

$$\mathcal{O} = \left\{ g\gamma g^{-1} \,|\, g \in PGL_n(\mathbb{C}) \right\} = \{ A \in sl_n(\mathbb{C}) \,|\, \operatorname{rank}(1+A) = 1 \}$$

is the set of traceless matrices A *such that* $1 + A$ *has* rank $= 1$.

Proof. Let $A = g\gamma g^{-1} \in \mathcal{O}$. Then

$$\operatorname{rank}(1+A) = \operatorname{rank}(g(1+\gamma)g^{-1}) = \operatorname{rank}(1+\gamma) = \operatorname{rank}\begin{pmatrix} 0 & & 0 \\ & \ddots & \\ 0 & & 0 \\ & & & n \end{pmatrix} = 1.$$

Conversely, if $\operatorname{rank}(1+A) = 1$, since

$$\operatorname{Tr}(1+A) = \operatorname{Tr}(1) = n,$$

we can find a matrix g such that

$$g^{-1}(1+A)g = \begin{pmatrix} 0 & & 0 \\ & \ddots & \\ 0 & & 0 \\ & & & n \end{pmatrix} = 1 + \gamma \qquad \Longrightarrow \qquad A = g\gamma g^{-1}. \qquad \square$$

Let us compute the dimension of $\mathcal{O}$. Let $A \in \mathcal{O}$. Then $\operatorname{rank}(1+A) = 1$, and so of the form

$$1 + A = \begin{pmatrix} a_1 b_1 & a_1 b_2 & \cdots & a_1 b_n \\ a_2 b_1 & a_2 b_2 & \cdots & a_2 b_n \\ \vdots & \vdots & \ddots & \vdots \\ a_n b_1 & a_n b_2 & \cdots & a_n b_n \end{pmatrix}.$$

Therefore A can be expressed as

$$A = \begin{pmatrix} a_1 b_1 - 1 & a_1 b_2 & \cdots & a_1 b_n \\ a_2 b_1 & a_2 b_2 - 1 & \cdots & a_2 b_n \\ \vdots & \vdots & \ddots & \vdots \\ a_n b_1 & a_n b_2 & \cdots & a_n b_n - 1 \end{pmatrix}.$$

The condition $\operatorname{Tr} A = 0$ gives the constraint

$$\sum_{i=1}^{n} a_i b_i = n.$$

There is an additional redundancy by rescaling

$$a_i \mapsto \lambda a_i, \qquad b_i \mapsto \lambda^{-1} b_i, \qquad \lambda \in \mathbb{C}^*.$$

It follows that $\dim_{\mathbb{C}}(\mathcal{O}) = 2n - 2$.

$$\mu^{-1}(\mathcal{O}) = \{(X, Y \in M \mid \operatorname{rank}([X, Y] + 1) = 1\}.$$

The **Calogero-Moser space** is defined to be the reduced phase space

$$CM_n = \mu^{-1}(\mathcal{O})/PGL_n(\mathbb{C}).$$

We have $(G = PGL_n(\mathbb{C}))$

$$\begin{aligned}
\dim_{\mathbb{C}} \mu^{-1}(\mathcal{O}) &= \dim_{\mathbb{C}} M - (\dim_{\mathbb{C}} sl_n(\mathbb{C}) - \dim_{\mathbb{C}} \mathcal{O}) = \dim_{\mathbb{C}} M + \dim_{\mathbb{C}} \mathcal{O} - \dim_{\mathbb{C}} G, \\
\dim_{\mathbb{C}} CM_n &= \dim_{\mathbb{C}} \mu^{-1}(\mathcal{O}) - \dim_{\mathbb{C}} G = \dim_{\mathbb{C}} M + \dim_{\mathbb{C}} \mathcal{O} - 2\dim_{\mathbb{C}} G \\
&= 2n^2 + (2n-2) - 2(n^2 - 1) = 2n.
\end{aligned}$$

We arrive at a $2n$-dimensional phase space CM_n.

4.5.2 Integrability of Calogero-Moser System

Next we show the Liouville integrability on this phase space. Consider the following functions $\mathscr{H}_i$ on M

$$\mathscr{H}_i(X, Y) = \operatorname{Tr} Y^i, \qquad i = 1, 2, \cdots, n.$$

It is clear that

$$\{\mathscr{H}_i, \mathscr{H}_j\} = 0, \qquad \forall i, j.$$

Since $\mathscr{H}_i$'s do not contain X. These functions are invariant under the $PGL_n(\mathbb{C})$-action, and so they give rise to n functions on the reduced phase space CM_n in involution.

Now consider a generic point $p \in CM_n$ represented by a pair of matrices (X, Y) such that

$$X = \operatorname{diag}(x_1, \cdots, x_n) \qquad \text{with } x_i \neq x_j \text{ for } i \neq j.$$

Let $Y = (y_{ij})$. Using $\mu(p) \in \mathcal{O}$,

$$\operatorname{rank}(1 + [X, Y]) = 1.$$

We have

$$\begin{aligned}
1 + [X, Y] &= \begin{pmatrix} 1 & (x_1 - x_2)y_{12} & \cdots & (x_1 - x_n)y_{1n} \\ (x_2 - x_1)y_{21} & 1 & \cdots & (x_2 - x_n)y_{2n} \\ \vdots & \vdots & \ddots & \vdots \\ (x_n - x_1)y_{n1} & (x_n - x_2)y_{n2} & \cdots & 1 \end{pmatrix} \\
&= \begin{pmatrix} a_1 b_1 & a_1 b_2 & \cdots & a_1 b_n \\ a_2 b_1 & a_2 b_2 & \cdots & a_2 b_n \\ \vdots & \vdots & \ddots & \vdots \\ a_n b_1 & a_n b_2 & \cdots & a_n b_n \end{pmatrix}.
\end{aligned}$$

- $a_i b_i = 1 \implies b_i = a_i^{-1}$.

$$1 + [X,Y] = \begin{pmatrix} a_1 & & \\ & \ddots & \\ & & a_n \end{pmatrix} \begin{pmatrix} 1 & 1 & \cdots & 1 \\ 1 & 1 & \cdots & 1 \\ \vdots & \vdots & \ddots & \vdots \\ 1 & 1 & \cdots & 1 \end{pmatrix} \begin{pmatrix} a_1^{-1} & & \\ & \ddots & \\ & & a_n^{-1} \end{pmatrix}.$$

By conjugating $\begin{pmatrix} a_1 & & \\ & \ddots & \\ & & a_n \end{pmatrix}$ on X, Y, we can set

$$a_1 = a_2 = \cdots = a_n.$$

Then

$$[X,Y] = \begin{pmatrix} 0 & 1 & 1 & \cdots & 1 \\ 1 & 0 & 1 & \cdots & 1 \\ 1 & 1 & 0 & \cdots & 1 \\ \vdots & \vdots & \vdots & \ddots & \vdots \\ 1 & 1 & 1 & \cdots & 0 \end{pmatrix} \implies y_{ij} = \frac{1}{x_i - x_j} \quad \text{for } i \neq j$$

and we denote $p_i = y_{ii}$. This representation of the point in CM_n is unique up to permutation of the diagonal elements of X. Precisely, let

$$\mathbb{C}^n_{\text{reg}} = \{(x_1, \cdots, x_n) \in \mathbb{C}^n \mid x_i \neq x_j\}.$$

Then there is an open embedding

$$T^*(\mathbb{C}^n_{\text{reg}}/S_n) \hookrightarrow CM_n$$

sending $(x_1, \cdots, x_n, p_1, \cdots, p_n) \mapsto X = \operatorname{diag}(x_1, \cdots, x_n)$,

$$Y = \begin{pmatrix} p_1 & \frac{1}{x_1 - x_2} & \cdots & \frac{1}{x_1 - x_n} \\ \frac{1}{x_2 - x_1} & p_2 & \cdots & \frac{1}{x_2 - x_n} \\ \vdots & \vdots & \ddots & \vdots \\ \frac{1}{x_n - x_1} & \frac{1}{x_n - x_2} & \cdots & p_n \end{pmatrix}.$$

Now we work on this open region $T^*(\mathbb{C}^n_{\text{reg}}/S_n)$. From the symplectic structure, we have the Poisson bracket

$$\{x_i, p_j\} = \delta_{ij}.$$

We have constructed n functions $\mathscr{H}_i = \operatorname{Tr} Y^i$ which are in involution. Let us consider

$$\mathscr{H}_2 = \operatorname{Tr} Y^2 = \sum_{i=1}^n p_i^2 - \sum_{i \neq j} \frac{1}{(x_i - x_j)^2}$$

which is precisely the Hamiltonian of the rational Calogero-Moser model for n particles on a line. It follows that $\mathscr{H}_1, \mathscr{H}_2, \cdots, \mathscr{H}_n$ gives n constants of motion in involution, showing that the system is Liouville integrable. The Hamiltonian flow of $\mathscr{H} = \operatorname{Tr} Y^2$ is

$$\rho_t(X, Y) = (X + 2tY, Y).$$

This is a motion of a free particle in the space of matrices. Representing it on $T^*\mathbb{C}^n_{\mathrm{reg}}$ produces solutions of rational Calogero-Moser system.

Bibliography

[1] Abraham, Ralph, and Jerrold E. Marsden. *Foundations of mechanics.* No. 364. American Mathematical Soc., 2008.

[2] Arnol'd, Vladimir Igorevich. *Mathematical methods of classical mechanics.* Vol. 60. Springer Science & Business Media, 2013.

[3] Arnol'd, Vladimir I., Alexander B. Givental, and Sergei P. Novikov. *Symplectic geometry.* Dynamical systems IV (2001): 1–138.

[4] Arutyunov, Gleb. *Elements of classical and quantum integrable systems.* Springer International Publishing, 2019.

[5] Babelon, Olivier, Denis Bernard, and Michel Talon. *Introduction to classical integrable systems.* Cambridge University Press, 2003.

[6] Born, Max, and Emil Wolf. *Principles of optics*, 7th Edition. Cambridge University Press, 2020.

[7] Calkin, Melvin G. *Lagrangian and Hamiltonian mechanics.* World Scientific, 1996.

[8] del Castillo, Gerardo F. Torres. *An Introduction to Hamiltonian Mechanics.* Springer, 2018.

[9] Cushman, Richard H., and Larry M. Bates. *Global aspects of classical integrable systems.* Vol. 94. Basel: Birkhäuser, 1997.

[10] Da Silva, Ana Cannas, and A. Cannas Da Salva. *Lectures on symplectic geometry.* Vol. 3575. Berlin: Springer, 2008.

[11] Etingof, Pavel. *Lectures on Calogero-Moser systems.* arXiv preprint math/0606233 (2006).

[12] Kibble, Tom, and Frank H. Berkshire. *Classical mechanics.* World Scientific, 2004.

[13] Landau, L. D., and E. M. Lifshitz. *Mechanics*, vol. 1 of Course on Theoretical Physics. American Journal of Physics 40, no. 7 (1972): 1050–1051.

[14] Marsden, Jerrold E., and Tudor S. Ratiu. *Introduction to mechanics and symmetry.* 1998.

[15] Milnor, John. *On the geometry of the Kepler problem.* The American Mathematical Monthly 90, no. 6 (1983): 353–365.

[16] Poole, Charles P., and Herbert GoldsteinJohn L. Safko. *Classical Mechanics.* 3rd Edition. John Wiley & Sons, 2012.

[17] Tong, David. *Classical dynamics.* Wilberforce Road, 2005.